North American Railroad Bridges

BY BRIAN SOLOMON

TO MANJIRI

This edition published in 2016 by
CRESTLINE
an imprint of Book Sales
a division of Quarto Publishing group USA Inc.
142 West 36th Street, 4th Floor
New York, New York 10018

First published in 2008 by Voyageur Press, an imprint of MBI Publishing Company LLC, Galtier Plaza, Suite 200, 380 Jackson Street, St. Paul, MN 55101 USA

Library of Congress Cataloging-in-Publication Data

Solomon, Brian, 1966–
North American railroad bridges / by Brian Solomon.
p. cm.
Includes bibliographical references and index.
ISBN: 978-0-7858-3391-8 (hardbound w/ jacket)
1. Railroad bridges—North America—Pictorial works. 2. Railroad bridges—North America—History. I. Title.
TG22.S65 2007
624.2—dc22
2007013998

Editor: Dennis Pernu
Designer: Sara Holle

Printed in China

Front cover: An excursion train crosses the former Soo Line bridge over the St. Croix River near Somerset, Wisconsin, on August 17, 1996. The bridge was built by Soo Line in 1910–1911 as part of a 17.5-mile cutoff. Each of the five cantilever arches spans 350 feet; the bridge, as built, was 2,682 feet long, and in 1911 the base of rail to the high-water mark was 169 feet. *Brian Solomon*

Frontispiece: On August 9, 1971, a pair of Vermont Railway Alco RS-3s pose atop the former Rutland Railroad bridge, known as the Brooksville Trestle, over the New Haven River. This is a Warren deck truss with overlapping sets of diagonals on each side for added strength. *Jim Shaughnessy*

Title pages: The viaduct spanning the valley of Tunkhannock Creek at Nicholson, Pennsylvania, is the Lackawanna's most enduring landmark. It is difficult to grasp the magnitude of this massive bridge when viewed from across the valley—that is until one sees a train roll across it at speed. In this view, the bridge silhouettes the setting sun on October 1, 1996. Each of its 10 main arches spans 180 feet; the bridge is 242 feet above its namesake creek. At 2,375 feet long and built for double track, it is generally cited as the largest reinforced concrete viaduct in the United States. *Brian Solomon*

Table of contents: The old Lackawanna main line, now used by Norfolk Southern to interchange freight with the Delaware-Lackawanna short line at Slateford Junction, Pennsylvania, crosses under the west end of the unused Delaware River Viaduct on the abandoned Slateford Cutoff. *Brian Solomon*

Back cover: Baltimore & Ohio engineer Albert Fink's bridges used cast-iron rods for compression members and wrought-iron tension rods. After 1880, most Fink trusses were replaced with heavier commercially designed iron or steel bridges. *U.S. Patent 10,887, May 9, 1854*

CONTENTS

Preface and Acknowledgments

In the Solomon family photo archives exists an image of myself in New Hampshire studying a disused, covered wooden railroad bridge along the old Boston & Maine. I was age two or three at the time. While I do not recall the details gleaned from that early interest, I do recall, about the same time, a photo excursion with my father, Richard Jay Solomon, on the New York subway system's Flushing Line, much of which rides an elevated structure across Queens.

In researching this text I've visited many bridges in the United States and abroad. I've also consulted dozens of books and hundreds of documents, communicated with a great many people, and inspected tens of thousands of images. The bibliography lists the majority of the written sources tapped for the project. Some of the most useful books for understanding the general history and function of bridges were Henry Petroski's *To Engineer Is Human* (1985) and *Engineer of Dreams* (1995), David B. Steinman and Sara Ruth Watson's *Bridges and Their Builders* (1957), David Plowden's well-researched and superbly illustrated *Bridges: The Spans of North America* (2002), and J. A. L. Waddell's epic *Bridge Engineering* (1916). Regarding the specifics of railroad engineering, I've long enjoyed William D. Middleton's works. His *Landmarks of the Iron Road* (1999) covers many of the great historical railroad spans in splendid detail. For those interested in the detailed history of Erie Railroad's Starrucca Viaduct, William S. Young's privately published *Starrucca: The Bridge of Stone* (2000) is essential reading.

On the topic of railroad bridges, so much has been written, any attempt to touch upon every aspect of the subject, let alone delve into every detail, is beyond impossible—the task would be so Herculean as to be unimaginable. One source consulted for this book suggested there were as many as 65,000 railroad bridges in the United States. How this figure was quantified is unknown, a fact that itself leads to a problem. By one definition, any span less than 20 feet counts not as a bridge, but as a culvert. Yet, not all writers have acknowledged this pedantic distinction. What I have chosen to include in this modest book are just a select handful of bridges relating to North American railroading that I found the most interesting and/or of significant historical relevance.

Among the challenges in sifting through the mountains of information accumulated for this project was reconciling incongruities and contradictory facts about individual bridges. Basic information such as dates of construction and dimensions have too frequently been reported differently by various sources. One author may simply overstate a bridge's statistics and the error gets perpetuated. Other situations have more specific difficulties: perhaps a bridge is modified during its lifetime and as a result lengthened or shortened. Somewhere along the line the old information is reprinted without qualification. I've made every attempt to cite individual sources for data and where I can find but one root source, to qualify statistics. In a few instances I cite seemingly contradictory statistics where I've found it impossible or impractical to determine which is more correct.

Many people have aided me in my research, photography, and production tasks. My father not only inspired my early interest in bridges through our trips together but helped in a great many other ways. As a child he crafted a model truss bridge from balsa wood, which fascinated me in my earliest years. Specific to the project, he provided key photographs, lent the use of his vast library of railway and engineering books, reviewed the text, and helped assemble the bibliography. Robert A. Buck of Tucker's Hobbies in Warren,

Massachusetts, inspired my interest in Major George Washington Whistler's achievements (and on numerous occasions brought me to visit Whistler's stone arches in the Berkshires), encouraged my interest in railways, and made enumerable introductions. Bob's wife, Silvia Buck, librarian of Warren Public Library, helped in my research of William Howe, himself late of Warren. Fairly late into the research for this text I realized that having grown up in western Massachusetts, I had been in the very bastion of "American trussdom," where more than a century earlier many of the foremost American truss patent holders had trod.

Special thanks to Patrick Yough, who made many connections for me, assisted with captions, on several occasions provided transportation and lodging, personally made first-class illustrations specifically for this project, and patiently waited during photography excursions while I inspected various spans. John Gruber assisted with research, wrote the sidebar on pontoon bridges, made introductions, and provided transportation and lodging in Wisconsin and California. John and I have traveled together many times over the last dozen years, looking at railways and railway bridges. Tom Farenz of the Iowa, Chicago & Eastern organized a visit to the swing bridge over the Mississippi at Rock Island, Illinois. Dave Abeles furthered my understanding of bridge engineering and helped review my text. John P. Hankey provided a tour of Baltimore bridges many years ago, leading me to and highlighting the importance of the Thomas and Carrollton viaducts and the last surviving Bollman truss. In addition, John answered innumerable questions, provided details on matters relating to the Baltimore & Ohio, and offered encouragement in my quest for knowledge. Niall Torpe patiently explained elements of bridge engineering and stresses while assisting with research on historic bridges in Britain and Ireland. Doug Riddell researched bridges on the Richmond, Fredericksburg & Potomac and Atlantic Coast Line, and provided extensive tours in Richmond, Virginia. Tom Carver organized trips on the Adirondack Scenic Railroad and helped with photography and photo captioning. Tom Hoover, who has accompanied me on trips since we were teenagers, brought me on my first visits to Pennsylvania Railroad's stone arches and enlightened me on matters relating to the PRR, B&O, Chesapeake & Ohio, and Clinchfield. William S. Young, whom I consider the foremost authority on the Starrucca and Tunkhannock viaducts, took time to share some of his research with John Gruber and me. Steve Smedley, an accomplished photographer, helped with captioning and participated in discussions on the distinctions between trestles and viaducts—a matter that has yet to be conclusively resolved despite definitive proclamations by pundits and iconoclasts. Keith Van Sant and Scott Snell helped locate bridges on the old Lackawanna. Tessa Bold provided lodging in Oxford, London, Washington, D.C., and Bonn, and has occasionally accompanied me on trips here and there. The Irish Railway Record Society in Dublin provided unlimited access to their archives, allowing me thousands of hours to read, research, and better understand bridges on both sides of the Atlantic. Hassard Stacpoole at the Association of Train Operation Companies in London helped me locate and visit historic bridges in England and Scotland. Thanks to the many railroaders, bridge tenders, signalmen, and engineers who have provided me access to facilities, answered questions, and bettered my knowledge of bridges over the years.

The text is but half of this book. Illustrations are the other half, and in this I also owe thanks to many people. I have been making photographs for more than three decades and some of my earliest attempts were of railroad bridges, but at no stage could I have done this without help. Many photographers, including those mentioned above, have traveled with me over the years: Brian L. Jennison, J. D. Schmid, Mel Patrick, Michael L. Gardner, George S. and Candace Pitarys, Tim Doherty, Mike Abalos, Dean Sauvola, Tom Danneman, Mike Danneman, Dick Gruber, Chris Burger, Scott Bontz, Blair Kooistra, Mark Hemphill, Brian Rutherford, Joe McMillan, Don Gulbrandsen, Joe Snopek, Dan Munson, Doug Moore, Doug Eisele, Mike Schafer, Otto Vondrak, Dave Burton, Don Marson, Gerald Hook, Danny Johnson, F. L. Becht, Ed Beaudette, George C. Corey, Howard Ande, Brandon Delaney, David Hegarty, Colm O'Callaghan, Mark Hodge, Paul Quinlan, Denis McCabe, Norman McAdams, John Cleary, Markku Pulkkinen, Mark Leppert, Bill Linley, George Melvin, Marshall Beecher, Pete Ruesch, Brian Plant, Vic Neves, Emile Tobenfeld, Will Holloway, Hal Miller, John Peters, and Norman Yellin. When traveling to make photos a portion of the vision is shared.

In reviewing photographs my goal was to find the most evocative and informative images. In addition to my own images, I've had the pleasure of reviewing thousands from other photographers and each is duly credited. Thanks to Tom Kline for access to the Lewis Raby archive and to Chris Guss for access to the Mike Abalos archive. Jim Shaughnessy graciously lent many of his own images as well as historic photos from his collection. Doug Eisele lent me both books and photographs from his collection as well as those of R. R. Richardson, Bill Dechau, and others. Scott Muskopf made images available from his collection. Special thanks to Otto Vondrak for producing bridge diagrams. I believe I've selected the most appropriate photographs to best illustrate the bridges in this project. The vast number of excellent photos left out of this book may be used to illustrate later books, while those published here will hopefully inspire photographers to make more and even better bridge illustrations.

My brother, Seán, lent me vintage postcards, traveled with me to inspect bridges, and unwittingly made connections regarding bridge engineers. My mother Maureen assisted with my complex travel logistics, international communication, and accounting.

Special thanks to Dennis Pernu and everyone at Voyageur Press for taking mere words and images and crafting them into this book.

Brian Solomon
March 2007

Introduction

Mankind has been constructing bridges for at least as long as we have found the time to write about them. The art of bridge building produced many fine examples before the advent of the railway. But the coming of the railway incurred a great need for new types of bridges. This spurred a frantic age of bridge building, as railways not only needed economically built bridges, but placed on bridge builders a host of new requirements, due to the great weights to be carried and obstacles to be crossed. Thus, railroads contributed to the rapid evolution of bridge design.

Traditional bridge designs were expanded and modified, new designs created, new materials employed, and new construction techniques developed and perfected. Certainly, railways were not alone in their need for better bridges, but a great many of the new bridge-building methods were used in the construction of railway bridges. In America alone, tens of thousands of railway bridges were built. As bridge builders learned their trade, the maximum spans (bridge-builder language for lengths) of railroad bridges increased dramatically. Perhaps as important, bridge builders learned to build with much greater economy while maintaining high levels of strength and stability, thus increasing safety. Nevertheless, the path to better bridge building was scarred with tragedy. The art and science of bridge building is still being perfected, and early on, some structures were swept away by floods, ice, wind, and other natural calamity; others collapsed as trains passed over them, sometimes sending railroaders and passengers to their doom.

As the American railroads grew, their bridge needs evolved. An important part of the story of American railroad bridge building has been the need for ever stronger, taller, and longer spans. In

On September 24, 1889, Cooperstown & Charlotte Valley No. 4 was the first locomotive to roll across the new bridge at Cooperstown Junction, New York. On either side of the iron lattice truss are three-tier timber frame trestles. *Photographer unknown, author collection*

The lattice truss at Proctorsville, Vermont, on the former Rutland Railroad is now used by the Green Mountain Railroad, a component of the Vermont Rail System, as seen on October 21, 2005. *Brian Solomon*

the early years of American railroading, railroads often needed to replace bridges as traffic demands required heavier locomotives and cars, greater operating speeds, and more tracks. Bridge replacement became an important part of bridge engineering. During the last decades of the nineteenth century and the first decade of the twentieth century, railroads replaced or rebuilt most of their bridges—often more than once. This effort, accompanied by line relocation and new construction, established much of the railroad infrastructure that exists to this day.

Advances in bridge design, along with new materials and efficient construction techniques, increasingly enabled railroads in North America and around the world to build bridges never before possible. Economical, solid bridges, combined with improved tunneling and modern railroad construction techniques, enabled railroads to be built rapidly and cost-effectively with private financing and little regard for most geographical impediments. Larger spans were erected and railroads reached places previously economically unattainable and unjustifiable. American railroads became characterized by tall steel viaducts, great steel and concrete arches, massive cantilever trusses, long tunnels, and deep rock cuttings.

When American railroads reached their zenith in the first decades of the twentieth century, they were among the largest and most important corporations in the world. Before the advent of modern highways, railroads dominated American business and transport. Without full command of the sciences of bridge engineering, fears of defective bridge design from the bad experiences of the previous century made bridge designers overly cautious. As a result of this conservatism, engineers tended to "overdesign" their structures. Bridges were made stronger than needed to provide a greater cushion of safety. Engineers built in a greater factor of safety than may

In 1914, Baltimore & Ohio performed much of the work on its new 12-mile Magnolia Cutoff in West Virginia on its Cumberland Division. The cutoff was designed to minimize the grade and greatly reduce curvature as compared with its old line, which hugged the south bank of the Potomac. Part of this construction included two heavy plate-girder bridges over the Potomac, one on either side of a tunnel through a ridge between Magnolia and Kessler's Curve. For many years, B&O operated both the old line and the cutoff as essentially a four-track line. In the 1960s, the old line was removed and now serves as a road. On September 19, 1997, a CSX freight rolls across the Magnolia Bridge. Beyond is the abandoned Warren deck truss on the old Western Maryland line that ran parallel to B&O's. *Brian Solomon*

have been necessary for axle loadings of the time, but they also took into consideration the likelihood of heavier axle weights in the future. The railroad industry has largely survived on this earlier conservatism, and many of the bridges installed during this period have proven adequate for modern times.

Railroad mileage peaked in the early 1920s, but by that time the industry had entered its long period of decline. As railroad finances gradually waned and public road building grew steadily, funds for new bridges became increasingly scarce. Railroads built few new lines and because of the great strength of their more modern structures had less need to replace bridges. While some magnificent new railroad bridges were constructed in the later years of the twentieth century—some as replacements, some on new routes and line relocations, and some in conjunction with highway projects—the vast majority of railroad bridges surviving today were constructed between 1890 and 1930.

68
68
MARC

CHAPTER 1

Early Masonry Arches

The masonry arch is one of the oldest and strongest bridge forms. The best known examples of early arched bridges are those built by the Romans, some of which still stand and are functional as aqueducts. Roman engineers perfected the use of stone, brick, and cement. (While the Romans also used wood for their bridges, those structures have long since vanished.) Among the finest examples of Roman arched stone viaducts is the Pont du Gard, at Nîmes, France, a three-tier arch-viaduct span over the River Gard.

Construction of massive and impressive masonry arches resumed during the latter Middle Ages and the Renaissance. In Britain, the famous London Bridge, credited to the design of Peter Colechurch, was begun in 1176 and completed 33 years later. It consisted of 19 pointed arches—some spanning just over 34 feet—and reached 936 feet, 6 inches across the Thames. This bridge was remarkable not just for its early date of construction and for its length, but because it was built across the rapidly flowing Thames, which required substantial piers to support the arches.

Left: Baltimore & Ohio's second significant bridge is the Thomas Viaduct at Relay, Maryland. Using eight elliptical arches, this 704-foot stone viaduct crosses Patapsco Creek on the railroad's line between Baltimore and Washington, D.C. Completed in 1835, the bridge is still in service today. In November 1995, a MARC (Maryland Rail Commuter Service) train crosses the bridge en route to Baltimore's Camden Station. *Brian Solomon*

Considered one of the world's most famous historical bridges, the old London Bridge survived more 600 years in daily service, only to be replaced at the dawn of the railway age in the early 1830s. By this time, the art of designing masonry arches was well established in Britain, and many fine examples were erected there in the early years of the nineteenth century to carry roads and as aqueducts for canals. Among the famous bridge engineers of this period and place were John Rennie and his son, Sir John Rennie; the latter designed a magnificent arched viaduct for the New London Bridge.

THE RAILROAD BRIDGE'S BRITISH ROOTS

The modern railway was born in Britain, where the first steam locomotives were built and refined. By the 1820s, the mode was further advanced as the first public "common carrier" railways were built for conveyance of freight and passengers. George Stephenson was among the most significant engineers of the period, building railways, locomotives, and bridges. Stephenson neither invented the locomotive nor the railway, but he used existing technologies in the construction of the world's first public steam railway: the Stockton & Darlington, begun in 1821 and completed in 1825. Prior to construction of this line, numerous industrial tramways had been built in England to serve collieries, quarries, and related industry. Although crude by comparison with the later public steam railways, some of the tramways used locomotives and required bridges. For example, about 1819, Josias Jesop engineered a five-span arched viaduct constructed of local stone for Mansfield & Paxton Railway.

Hunter Davis relates in Stephenson's biography that although the engineer considered a more modest bridge to cross the River Skerne (located east of Northgate, near Darlington) he was prevailed upon by the railway's directors to build a stone arched bridge that would "Look as impressive and imposing as possible." Stephenson contracted the Skerne cross to Ignatius Bonomi, a second-generation Italian Englishman working as an architect in Durham. Completed in 1825, the Skerne Bridge consisted of three stone arches, with the central span being the largest and most impressive. This bridge is considered the first railway bridge designed by a professional architect. It was prominently featured in a famous period painting by John Dobbin depicting the opening of the line that has become a symbol of the early railway, etching into the consciousness of railway pioneers the vision of a steam locomotive crossing a massive stone arch viaduct.

Stephenson employed masonry arches for later railway construction. For the Liverpool & Manchester Railway—a line that established important precedents for design, construction, and operation of British railways—Stephenson built the Sankey Viaduct. Described by Gordon Biddle and O. S. Nock in *The Railway Heritage of Britain* (1983) as the largest early railway viaduct, this bridge consisted of nine semicircular arches, each spanning 50 feet. At its highest point the viaduct is 60 feet above the River Sankey. Although George Stephenson and his son Robert also used innovative types of cast-iron bridges, the masonry arch would remain a standard type of British railway bridge throughout the nineteenth century. Numerous well documented examples of masonry arched bridges were built across Britain during the golden years of railway age.

Henry Law's *The Rudiments of Civil Engineering, 6th Ed.*, published in London in 1881, describes the masonry arch in great detail: "When an arch is in a perfect state of equilibrium, if we suppose its abutments incapable of yielding, it can only fail in consequence of the crushing of its material, the cohesive power of which is then the limit of the strength of the arch."

The basic features of a simple masonry arch are the abutments, located at both ends of the bridge and serving as the portions of the substructure that sustain the natural outward thrust of the arch, and the arch ring, typically comprising *voussoirs*, wedge-shaped stones or bricks fitted together and capped at the crown by a keystone. The interior surface of the arch is known as the *intrados*; the exterior curve of the arch is the *extrados*. The intrados meets the abutment at the *springing*. The *span* of the arch describes the horizontal distance bridged, while its vertical measurement is the *rise*. Often, arches are linked to form a series of spans, such as the Pont du Gard and London Bridge described earlier and a great many early British railway viaducts. While arched bridges have been built using a variety of arch profiles,

Isambard Kingdom Brunel's Sounding Arch at Maidenhead was by far the broadest and lowest arch bridge of its day. Built in 1839 of yellow brick, it was widened by Sir John Fowler between 1890 and 1893, using red bricks, to accommodate four tracks. On March 19, 2007, a First Great Western HST (high-speed train) passes over the famous bridge that crosses the River Thames a mile east of Maidenhead, England. *Brian Solomon*

including elliptical and pointed, the majority have used semicircular arches.

Early railway engineers soon learned to push the traditional arch to new limits. Isambard Kingdom Brunel, the visionary engineer behind the Great Western Railway, designed and erected an arched bridge made of brick and using an unprecedented broad curve across the Thames at Maidenhead near Reading in 1835. Known as Brunel's Sounding Arch, this bridge crosses the river using a pair of slightly pointed elliptical arches, each spanning 128 feet with a rise of just 24 feet, 4 inches and a curvature at its crown of 169 feet—by far the broadest radius attempted with a masonry arch at that time. Approaches to the main spans on each side of the river consist of more conservative semicircular arches. One of the iconic symbols of the age, the Sounding Arch was depicted in 1844 by J. M. W. Turner in one of his most famous paintings, *Rain, Steam, and Speed—The Great Western Railway*. Henry Law wrote, "[The Maidenhead Bridge] is certainly the boldest which has ever been constructed, the actual pressure at the crown of the arch being about one-third of that which begin to injure the cohesive strength of the material of which it is composed." At the time of Law's comment, the bridge had been in regular service for more than four decades. Although widened to accommodate four main tracks, the bridge today remains on the Great Western main line.

The advantages of the masonry arch were obvious to early railroad builders. This time-tested form is exceptionally strong and thus capable of withstanding the weight of locomotives and the forces they place upon the bridge as they cross at speed. A well-constructed arch is as strong as the individual stones that are used in the structure. The methods for erecting masonry arches from either stone blocks or brick were well-known, and, aside from keeping the drains open, the masonry arch requires a minimum of maintenance.

Additionally, the historic connotations of stone arch bridges were not lost on the early railway builders. As pioneers of a new enterprise, privately funded but in need of public acceptance, railway builders desired to inspire confidence in their investors. Railways needed the appearance of being well-built and exuding permanence for the ages. Early railways were viewed as new corridors of commerce to be enjoyed for generations. The solid quality of magnificent stone arches lent both physical and psychological support to the railway in its formative stage of development.

Yet for all their strength and durability, there were distinct disadvantages in building masonry arches, as America's early railroad builders soon learned. Stone arches required lots of time to build and entailed exceptional cost. Also, masonry bridges tended to greatly exceed the loading requirements of the day, so stone arched bridges were overbuilt for the traffic they were intended to carry. Many masonry arches have survived in service despite significant increases in the weight and speed of trains.

AMERICAN RAILWAY ARCHES

Considering that the Atlantic Ocean separates America and Britain and that it took weeks to cross in the early nineteenth century, it is fascinating how much knowledge was transferred between these two countries in that period. Notably, America followed Britain's lead in development of inland waterways; in 1825, New York State completed its famous Erie Canal, which opened a through-transportation corridor from New York City to the Great Lakes. Among the first serious railway projects in the United States was that put forward by the Delaware & Hudson Canal Company, which in 1827 began building a rail-tramway for the conveyance of coal from its mines to its canal at Honesdale, Pennsylvania. The same year, the Baltimore & Ohio Rail Road was chartered to compete with the Erie Canal by means of a common carrier railroad to funnel traffic from Baltimore through the Allegheny Mountains to Wheeling, Virginia (today in West Virginia), on the Ohio River and thus to the nation's interior. B&O's promoters were very much aware of George Stephenson's recent achievements and no doubt hoped to emulate his success in Maryland.

Significantly, both D&H and B&O dispatched engineers to Britain to study railway technology, expertise, and materials. In 1828, D&H sent Horatio Allen, a 27-year old engineer who acquired more than just knowledge. America's industry was then in its formative stages and not yet ready to produce railway technology commercially, so Allen ordered iron rails and four complete locomotives with which to build and operate D&H's proposed line. B&O followed a similar quest a year later, sending several young engineers, including Jonathan Knight, Ross Winans, William G. McNeill, and George Washington Whistler to Britain. McNeill and Whistler were graduates of the United States Military Academy at West Point, one of the first significant institutions for educating civil engineers in America. They met with the great railway builders

Family Affairs

The engineering talent pool in the early years of railroading was small and closely knit, with key figures often related to one another. As noted in this chapter, both the Rennies and Stephensons were father-and-son partnerships in Britain. Later, in America, the Roeblings, known for construction of suspension bridges, especially the famed Brooklyn Bridge, followed this pattern. William G. McNeill and George Washington Whistler were related through marriage—McNeill's sister Anna was Whistler's second wife. (Incidentally, their son, James McNeill Whistler, became the famous American painter whose best-known work, *Arrangement in Grey and Black No. 1*—better known as *Whistler's Mother*—is displayed at the Musée d'Orsay in Paris, formerly the railway station Gare d'Orsay.) Typically working together, Whistler and McNeill were among America's first railroad boomers—a traditional railroad term normally applied to skilled operating men who followed work by moving from carrier to carrier as traffic "boomed" and created employment opportunities. They worked for B&O in its formative years, but had moved on to engineering projects for Baltimore & Susquehanna in 1830 and the Paterson & Hudson River in 1831. In 1832 they migrated to New England where they built the first public railways in that region. These lines, like the B&O, served as proving grounds for the next generation of American railway engineers.

America's first significant railroad bridge is Baltimore & Ohio's Carrollton Viaduct, built in 1829. This November 1995 view depicts the historic stone bridge, which closely resembles British-built structures from the period. The main arch consists of an 80-foot span and a 58-foot rise over Maryland's Gwynns Falls, located a short distance from B&O's original Baltimore terminus. Carrollton Viaduct is sometimes heralded as America's first railroad bridge–which is true, if one ignores a nearby stone culvert of a roughly 20-foot span that was built just prior. *Brian Solomon*

George and Robert Stephenson, who were then nearing the zenith of their careers, and legendary bridge builder Thomas Telford. It is of little surprise that B&O took its queues from British practices established by Stephenson and others; among the other precedents set by Stephenson was the track gauge of 4 feet, 8.5 inches adopted by B&O and many other formative lines in the United States. Although it wasn't standardized until the 1880s, the Stephenson gauge is now the American standard, as well as the British, and is the most widely used track gauge around the world.

The B&O's Carrollton and Thomas Viaducts

On July 4, 1828, B&O initiated westward construction. Significantly, the contract for construction of its first large bridge to cross Gwynns Falls (only a short distance from B&O's first Baltimore terminus) was given to established bridge builder Caspar W. Wever in December 1828. (Wever later became B&O's construction superintendent.) Bold projects are rarely accomplished with ease. Although visions of the Stockton & Darlington's Skerne Bridge and significant British bridges such as the Causey Arch at Tanfield resonated with B&O's pioneers, proponents of an arched bridge across Gwynns Falls needed to overcome personal preferences before the arch could take shape. While Wever envisioned a magnificent stone arch of classical design, other B&O engineers preferred a wooden span, as discussed in *History of the Baltimore & Ohio Railroad* by John F. Stover (1987), and *Landmarks of the Iron Road* (1999) by William D. Middleton. Wever's opponents argued that a wooden bridge would be cheaper to build. Among them was Colonel Stephen H. Long, then serving as a B&O engineer and known for a patented type of wooden truss bridge popular on American railways in the formative years of construction. Despite opposition, Wever's philosophy prevailed and he built his stone arch bridge with the help of his experienced assistants, James Lloyd and John McCartney. Construction began in May 1829 and was completed by December. The bridge's primary span was an 80-foot arch over the river, with a secondary 20-foot arch over a wagon road. The total length of the bridge was 300 feet. It was 26 feet, 6 inches wide, more than adequate for carrying a double-track line. At its highest point it was 58 feet above the river. Named for Charles Carroll of Carrollton, Maryland, this bridge has been known since its dedication as the Carrollton Viaduct. So exciting was the completion of the arch, some of America's first public passenger trains were the excursion trains organized to take people 1 mile from B&O's Mount Clare Depot to see the new viaduct in December 1829. While Wever succeeded in building this bridge, his original cost estimates were miscalculated by a factor of nearly four.

Although the B&O constructed a new main line to Baltimore in 1868, bypassing its original route, the Carrollton Viaduct remained active on a secondary line and today carries CSX tracks.

A few years after it began construction, the B&O decided that a branch toward Washington, D.C., would be a prudent investment. At Relay, Maryland (so named for the station where horses were exchanged in days before B&O used locomotives), near where the branch diverged from the main line, is the Thomas Viaduct, designed and built by Benjamin H. Latrobe, son of famous architect and engineer Benjamin Latrobe, a British immigrant who was educated in architecture and among the most respected men of his generation. He taught his son the principles of his trade, conveying appreciation for classical designs and the great architectural achievements in Britain and Continental Europe.

Latrobe the younger had worked closely with B&O's Jonathan Knight and was familiar with the great British bridge builders. His 704-foot-long curved stone Thomas Viaduct crosses Patapsco Creek 66 feet above river level on eight elliptical arches. The curve is a relatively severe 4 degrees. Construction of the viaduct began in August 1833 and was completed by July 4, 1835. It was named for Philip E. Thomas, B&O's first president. Today it carries both CSX freight trains and Maryland Rail Commuter Service (MARC) suburban passenger trains.

Remarkable because of their early construction and exceptional longevity and durability, the great cost of these early viaducts led B&O to find more cost-effective methods of building bridges. B&O engineers would thus pioneer iron truss designs, discussed in the next chapter. However, most of the early trusses are long gone, while most of the first stone arch bridges are still in use.

The Canton Viaduct

New England boasts America's earliest commercial railway project: the Granite Railway of Massachusetts at Quincy, conceived in 1823 and built in 1826 to move granite for the construction of the Bunker Hill Monument. In the 1820s, New England was among the most prosperous and most industrialized regions of the United States, and thus had considerable need for improved transport. Three early public railroads along the lines of the B&O were underway in the first half of the 1830s. Radiating from Boston were the Boston & Lowell, Boston & Worcester, and Boston & Providence. These lines, along with other lines such as the Norwich & New London (soon renamed the Norwich & Worcester) variously took advantage of McNeill's and Whistler's skills during their construction.

Although not the first line to lay track in Massachusetts, the B&P was the first to require a substantial bridge. Southwest of Boston, B&P's line crossed the valley of the Neponset River at Canton. As with the B&O's Thomas Viaduct, it was impractical for the railway to drop down into the valley, cross the river on a smaller bridge, then climb back out again. To do so would require unreasonably steep gradients, making operation with locomotives impossible. In order to maintain a relatively even gradient, McNeill planned a huge stone viaduct to cross the valley and the river. McNeill's Canton Viaduct is a distinctive design, somewhat resembling a tall and buttressed stone wall with small portals at its base rather than a traditional arched viaduct. As with the Thomas Viaduct, which was under construction about the same time, this bridge was laid out on a curve, although a much gentler one.

In *Landmarks of the Iron Road*, Middleton cites the Canton Viaduct's dimensions as spanning 615 feet and featuring six small semicircular arches at its base, each spanning 8 feet, 4 inches to allow passage of the Neponset. A larger 22-foot arch at the

The Western Railroad of Massachusetts was among the most significant American civil engineering projects of its day. Its great need for bridges not only produced classic stone arches, but inspired local engineers to advance truss design. The first application of the Howe truss was just a few miles east of this elliptical stone arch bridge, which has carried trains over the Quaboag River at West Warren since 1840. In 1987, a loaded Conrail autorack train destined for Westborough, Massachusetts, works eastward upgrade through the Quaboag River Valley. *Brian Solomon*

Providence end accommodates a road through the valley. The viaduct consists essentially of two parallel walls, each roughly 5 feet wide and hollow inside. As built, the viaduct was 22 feet wide at the top. Construction began in April 1834 and was completed at the end of July 1835. Remarkably, this unusual bridge has survived in regular service ever since and today is part of the Northeast Corridor—the electrified high-speed main line between Boston, New York, and Washington, D.C. Over the years, the Canton Viaduct has been reinforced and modified to accommodate double track, heavier, faster trains, and most recently catenary supports for electrification.

The Western Railroad Arches

Perhaps the most significant early railway in Massachusetts was the Western Railroad that was intended to connect with lines in New York State and provide an all-rail route between the capitals of New York and Massachusetts. As early as 1829, railroad proponents had seriously discussed construction of a line from Boston to Albany to tap Erie Canal traffic. The Boston & Worcester was the first leg of this vision, chartered in 1831 and completing 44 miles to Worcester by 1835.

The Western Railroad of Massachusetts was chartered on March 15, 1833, and like the other early New England lines, was engineered by Whistler and McNeill, although Whistler played a more significant role in the Western. Between Worcester and Springfield, the Western crested Charlton Hill at an elevation of 907 feet, 9 inches above sea level. This required a significant grade, especially on the descent to East Brookfield on the west slope. The bridges originally built on this portion of the line were largely wooden trusses of the type promoted by Colonel Long (Warren was the site of the first Howe truss, and the largest bridge on the line was the wooden truss span across the Connecticut River at Springfield). A stone, double-span arch bridge at West Warren presumably was designed by Whistler and his engineering team, but very little is known of it.

West of the Connecticut River was Whistler's greatest engineering challenge: the Berkshires. In

Neatly situated in the Quaboag River Valley west of West Warren, Massachusetts, is this double-elliptical arch bridge usually credited to Major George Washington Whistler, who built the Western Railroad in the late 1830s. Although Whistler's larger stone bridges in the Berkshires are well documented, scant information has been found relating to this structure. On May 8, 1997, an eastward Conrail autorack train glides over the bridge on its way up the valley. *Brian Solomon*

When New York Central implemented major improvements to the Boston & Albany between 1911 and 1912, it relocated more than a mile of line on the east slope of Washington Hill, from east of milepost 129 to just west of milepost 130. In so doing, it isolated three of the massive masonry arches built by Alexander Birnie for George W. Whistler during 1840–1841. This arch on the old alignment is readily visible from the 1912 alignment across the West Branch of the Westfield River. *Brian Solomon*

surveying and constructing the Western over the Berkshires, Whistler completed the world's first substantial railway over a mountain grade relying strictly on adhesion. He built the whole line to accommodate double-track, demonstrating considerable foresight, since it would be a number of years before traffic volume would warrant two main tracks. The most difficult part of the Berkshire grade was the east slope of Washington Hill. Here, Whistler surveyed a sinuous alignment that crossed back and forth across the West Branch of the Westfield River (the English name for the Pontoosuc). In 1843, *American Railroad Journal and Mechanics' Magazine* reprinted extracts from the "Report of the Western Railroad," which detailed the construction of the line: To cross the river, Whistler designed 10 masonry arch bridges. The smallest of these had a span of 15 feet, while the three largest spanned 60 feet (some sources indicate 70 feet). The height of the bridges ranged from 12 feet to 67 feet. Whistler also designed two stone arches to carry the line over roads. Most of the arches were built over the tortuous 3-mile section near Middlefield Station. In addition to the arch bridges, substantial retaining walls and related construction were required. "Report of the Western Railroad" estimated that 220,586 "perches" of stone (an old measurement representing 25 cubic feet) were used in construction of the line, much of which was for the bridges.

Grading was completed on Washington Hill by the end of 1840, and the line was opened to traffic

Whistler's ascent of the east slope of Washington Hill was among the most challenging railroad construction projects of its day. Whistler laid out a sinuous climb whereby the Western Railroad crossed and recrossed the West Branch of the Westfield River more than a dozen times in just a few miles. To bridge the river, he designed a number of exceptionally durable stone arches and contracted construction to Alexander Birnie. Only a couple of the original bridges remain in service, although three, such as this one west of milepost 129, survive unused on an alignment abandoned in 1912 when New York Central improved the line. The others were destroyed many years ago. *Brian Solomon*

Detail view of Whistler's double stone arch west of milepost 128 near Chester, Massachusetts. This is one of two surviving stone arches that remain in service on the east slope of Washington Hill. Today, this bridge carries CSX's Boston Line, which hosts as many as 30 trains daily. *Brian Solomon*

in May 1841, although some of the bridges were not finished until 1844. CSX now operates the line; west of the Connecticut River two of these massive masonry arches remain in service.

Starrucca Viaduct

The last significant early masonry bridge was the Starrucca Viaduct built by the New York & Lake Erie (Erie). Its story is closely linked to the earlier masonry arches. The Erie was another early ambitious railroad endeavor, and like the B&O and B&W/Western projects, it was first envisioned in the 1820s. Unlike the B&O and New England roads, the Erie did not adopt Stephenson's track gauge, but instead used broad-gauge tracks measuring 6 feet between the inside rail flanges—the widest adopted by any common carrier in the United States. Promoted by New York interests, the Erie originally was chartered to run from the lower Hudson Valley at Piermont, New York, to the Lake Erie port town of Dunkirk (west of Buffalo) and remain entirely within the state. The Erie was slow to get started; ground wasn't broken until 1835, and heavy construction wasn't underway until the mid-1840s. By this time the route had varied from its early plan, and at several points the railroad crossed into Pennsylvania, which was necessary to maintain the shortest course with reasonable gradients and construction costs. Beyond Gulf Summit, New York, where the railroad crossed the divide between the valleys of the Delaware and Susquehanna rivers, the line dropped into Lanesboro, Pennsylvania, on its way to Binghamton, New York. Here it was surveyed to cross an especially picturesque valley over Starrucca Creek.

Deep in the Valley of the Pontoosuc–now known as the West Branch of the Westfield River–a westward Conrail train grinds up the west slope of Washington Hill, crossing Whistler's double stone arch west of milepost 128 on the afternoon of November 13, 1992. *Brian Solomon*

Starrucca's masonry arches catch the light of the morning sun in October 2001. Depending on how one qualifies construction expenditure, in 1848, Starrucca Viaduct cost the Erie between $310,000 and $336,000, making it among the most expensive individual pieces of railroad infrastructure at that time. Today, it is hard to imagine men erecting Starrucca's classical masonry arches. After more than 150 years, they seem a natural part of the valley at Lanesboro, Pennsylvania, as if they have stood there forever. *Brian Solomon*

In his self-published book, *Starrucca: The Bridge of Stone, Sesquicentennial Edition* (2000), William S. Young, in addition to providing detailed descriptions of the bridge and its construction, highlights the complex familial relationships between the Erie's various engineers and their relationships to other significant personalities of the period. Tackling the task of crossing the valley was Erie's Julius W. Adams, a man of considerable talent. Adams, by no coincidence, was Whistler's nephew, and instrumental in building the viaduct was Adams' brother-in-law, James Pugh Kirkwood, a Scottish immigrant. Both had worked for Whistler and McNeill on their New England projects and were familiar with the masonry bridges on those lines. Later, Adams and Kirkwood helped found the American Society of Civil Engineers. Adams designed the tall masonry arches of Starrucca Viaduct, while Kirkwood oversaw construction, which began in 1847 and concluded on November 23, 1848.

Also involved with the construction of the Erie was Horatio Allen, the pioneering American railroad builder mentioned earlier, who served as the

Erie's consulting engineer and briefly as its president. According to Young, Allen—who had observed some of the pioneer masonry construction in Britain, including the Sankey Viaduct—reviewed Adams' plans for Starrucca.

Starrucca consists of tapered stone piers measuring up to 65.67 feet from their foundations to arch

In the 1980s, the old Erie route was Conrail's preferred means of moving double-stack container trains east of Buffalo, New York. Three of the railroad's new General Electric DASH 8-40Bs haul more than a mile of American President Lines containers eastward over Starrucca Viaduct in April 1989. The bridge's solid arches are capable of supporting the weight of modern trains, which are considerably heavier than those of the 1840s, when the bridge was designed. *Brian Solomon*

This view, circa 1900–1910, shows an Erie Railroad camelback steam locomotive leading an eastward freight across Starrucca. In Erie days, this picturesque viaduct was officially identified as bridge 189.46, denoting the mileage measured from the railroad's terminus at Jersey City. At the time this image was made, the Erie was a primary corridor for commerce. Today, the Erie route over Starrucca survives as a lightly used secondary railroad line.
Jim Shaughnessy collection

springings. They were erected across the valley using timber falsework. These piers support 17 semicircular stone arches, each with a span of 51 feet and a rise of 20 feet. The bridge is on the westward ascent of Gulf Summit, rising on a steady 1.2 percent grade, and so Starrucca is lower at its west end than at the east. At its highest point, Starrucca Viaduct is 100 feet above the river. It is 1,040 feet long and 25 feet wide at its top. Starrucca has endured almost 160 years of change on the Erie line. In 2005, Susquehanna's parent company, Delaware Otsego, took over from Norfolk Southern the operation of the old Erie route east of Binghamton and typically operates three trains each way per week.

Several decades would pass before American railroads developed a renewed interest in masonry construction on a large scale, and the majority of the later masonry bridges were designed as replacement structures (discussed later in this book). In an interview, historian John Hankey pointed out a few interesting exceptions to this trend, specifically, the elliptical arch bridges built by Central of Georgia. One of these, located at Savannah, dates from 1859 and is a rare example of antebellum masonry railroad construction that survives to the present day.

Amtrak 576, one of 140 EMD SDP40Fs that comprised the passenger system's first motive-power purchase, works its way across the Stone Arch Bridge in Minneapolis, Minnesota, on March 1, 1975. Built in 1883 by rail baron James J. Hill, the bridge is the only span of its type across the Mississippi River. Measuring 2,176 feet in length, its 23 remaining arches are made of native granite and limestone. In 1963, two of the original arches were replaced by the truss at left, which spans an Army Corps of Engineers navigation channel. The Stone Arch Bridge served as a railroad bridge until 1978; today it serves pedestrians, bicyclists, and a trolley. *John Gruber*

In the formative years of railroad construction, covered bridges were common. But by the early 1900s, with advances in bridge design and the trend toward heavier axle loading, combined with the desire for permanence, most of the old covered railroad bridges were replaced. This type survived in a few places, notably northern New England. The last in service was a covered lattice truss known as the Fisher Bridge near Wolcott, Vermont, which used a combination of the Pratt and Town styles. In February 1970, a St. Johnsbury & Lamoille County freight led by a pair of GP9s crosses a venerable structure that was completed about 1908. The bridge survives at the time of this writing. It was last used for tourist excursions in the 1990s, but the line has since been abandoned. *Jim Shaughnessy*

CHAPTER 2

Truss Bridges

The truss is one of the most basic bridge types, one that has been well adapted to railroad applications since the early days of American railroad building.

EARLY WOODEN TRUSS TYPES

The first practical development of the truss span is usually credited to Italian engineer Andrea Palladio, who in 1520 published detailed descriptions of four different truss-bridge arrangements. While a number of wooden trusses were constructed in Europe as a result of Palladio's work, the design was rarely used for new construction for more than two centuries. Then, after a considerable hiatus, in the latter part of the eighteenth century there was renewed interest in European truss-bridge construction.

Truss construction was familiar to American builders who commonly erected truss roofs. American truss designers made use of timber construction; a logical choice, considering that in North America there was a great quantity of good-quality timber and a labor pool familiar with woodworking. In the early years of the nineteenth century, wooden covered bridges grew in favor in the United States, resulting in a variety of new timber truss designs. Among the most significant was Theodore Burr's covered bridge

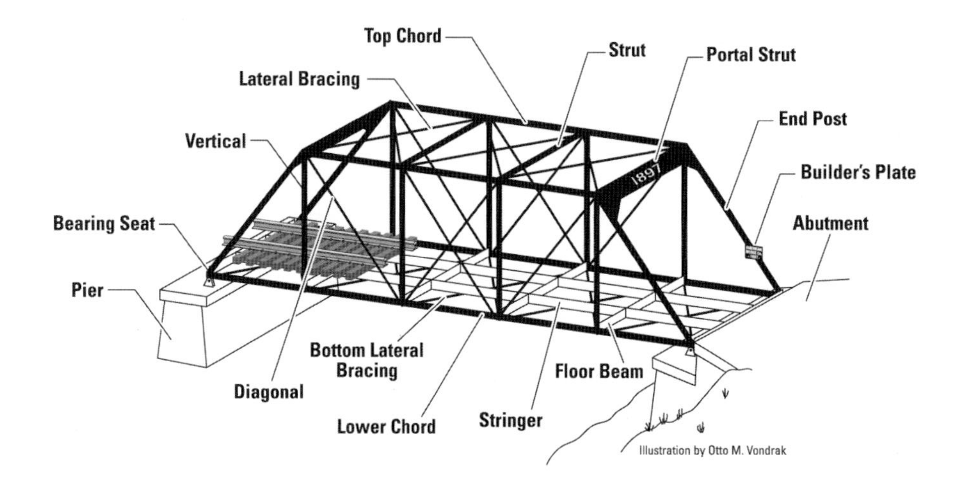

supported upon its abutments by vertical force without imposing any horizontal thrust, as do arched trusses. In *The Early Years of Modern Civil Engineering* (1932), Richard Shelton Kirby and Philip Gustave Laurson wrote that "Town's lattice trusses were known the world over and were the prototype of many later iron bridges. The indebtedness of subsequent designers to

using an arched truss arrangement. This blended the basic forms of girder and arched bridges, with the arch adding strength to the truss structure.

In 1820, architect and bridge builder Ithiel Town of Connecticut patented what he called a "lattice truss." This involved numerous parallel and diagonal supports between upper and lower chords in a multiple-intersection crisscross pattern carrying the stress of the bridge in the form of girder. This design is sometimes described as a "true truss" because it is self-supporting and not reliant upon an arch or other form of exterior bracing beyond its basic truss, which includes pieces, or "members," that are engineered to balance the opposite forces of tension and compression inherent in bridge design. Thanks to that balance, the true truss is

Town was rather generally recognized and admitted, which is not always the case in such matters." In contrast, David Plowden's more current analysis, *Bridges: The Spans of North America* (2002), dismisses the importance of Town's truss to later metal lattice trusses, explaining that the metal designs were of European origin.

Town's significance was his method of carrying the stresses of the bridge using numerous diagonals that allowed individual members to be cut from lighter and less substantial timbers. Although initially applied to road bridges—among them the many classic covered bridges once emblematic of rural America—Town later adapted his design for heavier applications, including railroads, receiving a revised patent in 1835. This was accomplished, as

This interior view of a Town lattice-style truss at an unknown location in Vermont's Lamoille Valley clearly shows the method of construction and the multiple-intersection diagonal members. Among the advantages of the Town truss was that individual members could be fashioned from lighter and less substantial timbers. Also, the bridges were relatively easy to assemble without special tools or skills. However, like all wooden bridges, they were susceptible to fire and easily destroyed. *Lewis Raby photograph, Tom Kline collection*

A pair of 4-4-0 steam locomotives tests the strength of the Boston, Concord & Montreal bridge over the Connecticut River between Wells River, Vermont, and Woodsville, New Hampshire. A double-deck wooden arched truss carried the road on the lower level and the railroad on top. Notice how the arch reaches to the bottom of the abutment. The BC&M later was part of the Boston & Maine system. In 1903, a pin-connect Pratt deck truss was built here, and the road relocated on a new alignment. *Author collection*

Boston & Maine's pin-connected Pratt deck truss between Wells River, Vermont, and Woodsville, New Hampshire, was constructed in 1903 as a replacement for an earlier bridge over the Connecticut River. Ninety years later, it was still in service. In this October 1993 view, a short freight is led by New Hampshire & Vermont Railroad GP9 No. 669 eastward over the bridge on its way from White River Junction, Vermont, to Whitefield, New Hampshire. The line has since been abandoned. The depth of a truss is a function of the length of the span and the forces of live load it is intended to carry. The greater the load and the longer the span, the deeper the truss needs to be. *Brian Solomon*

Plowden describes, by doubling the lattice web. Kirby and Laurson cite several prominent examples of Town truss railroad bridges. Philadelphia & Reading used the design prior to 1839, making these among the earliest railroad truss bridges on record, but perhaps the most significant Town railroad bridge was Richmond & Petersburg's 2,900-foot-long, multispan crossing of the James River at Richmond, Virginia. (The R&P was a predecessor of the Atlantic Coast Line, one of the railroads that became today's CSX.) The James River Bridge required 18 masonry piers with individual spans as long as 153 feet. Kirby and Laurson quote W. Hosking's *Essay and Practical Treatises on the Theory and Architecture of Bridges* (1843): "This bridge was completed in September, 1838, having occupied less than two years in its construction, its cost amounting to about £24,200 sterling. We doubt whether any part of the world could produce an instance of a work equal in magnitude to this being executed for so small a sum."

This last comment is key to understanding why the truss was so widely adopted by American railroads: ease of construction, combined with very low cost of transporting components to the construction site, gave timber trusses a great advantage over the far more costly masonry arch for long-span bridges. In the formative period of American railroading, vision typically outpaced finances. As a result, railroads needed to make the most of what little funding they could drum up. A cheap bridge, even if viewed as a temporary structure, had great initial appeal over a very costly, albeit better built and more permanent, bridge. The Town truss required a comparatively small amount of timber and could

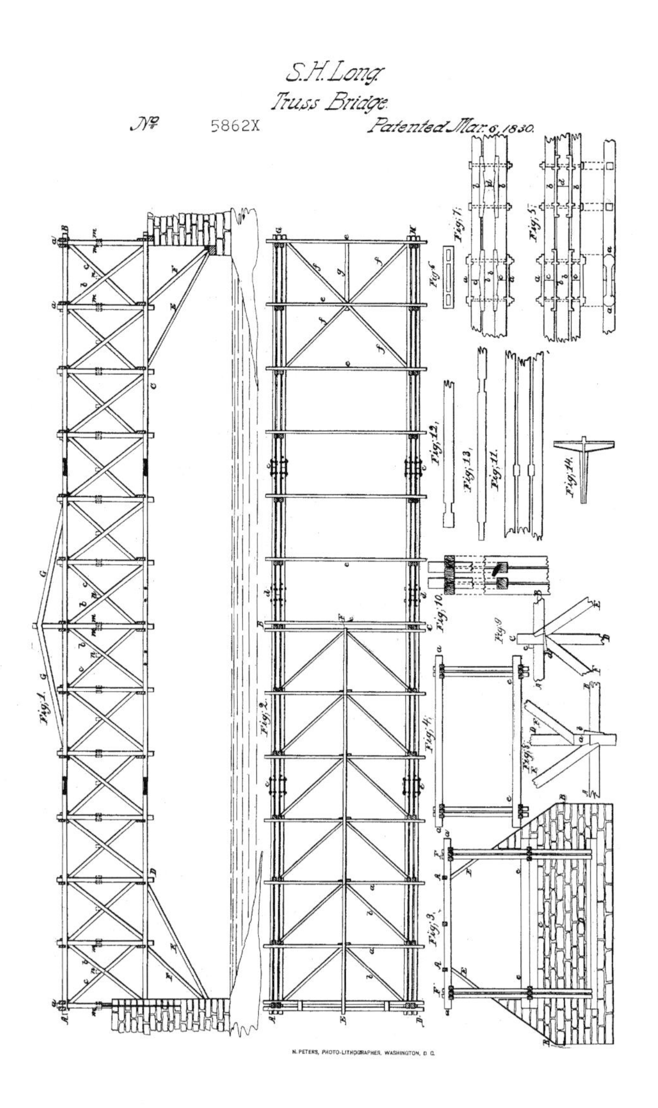

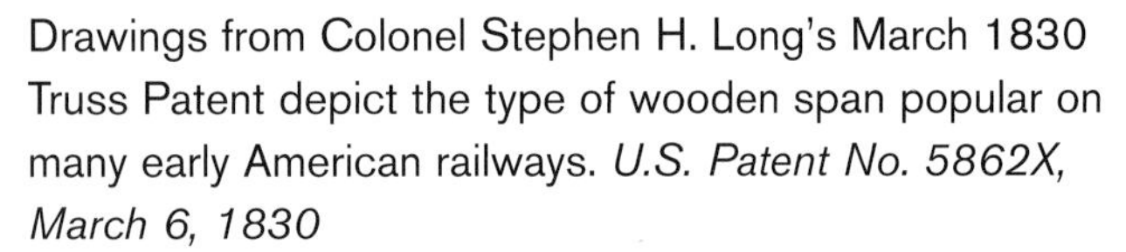
Drawings from Colonel Stephen H. Long's March 1830 Truss Patent depict the type of wooden span popular on many early American railways. *U.S. Patent No. 5862X, March 6, 1830*

be erected quickly and relatively easily. Despite a multitude of advancements in bridge design, the simplicity and economy of the Town truss continued to find proponents through the nineteenth century. Walter M. Macdougall's book, *The Old Somerset Railroad* (2000), describes in detail Town's lattice-covered bridges built on the Somerset in rural Maine in the 1870s.

The demonstrated success of Town's timber truss, combined with the growing need for bridges (especially railroad bridges), resulted in a plethora of new truss designs that were patented by American engineers during the formative years of railroad construction. Baltimore & Ohio's Colonel Stephen H. Long patented a wooden panel truss design in 1830 that was much used by railroads through the early 1840s. Kirby and Laurson considered Long's patent a predecessor to the Warren and Pratt truss designs widely adopted in the latter part of the nineteenth century.

ENTER WILLIAM HOWE

Central Massachusetts was a hotbed of creativity for bridge engineers in the 1840s and 1850s. A surprising cluster of patents were granted to various inventors living in towns between Worcester and Springfield during this period. Most of the patents were for esoteric designs with little practical application. However, one of the earliest, and by far the most significant, truss designs that originated in this region was the work of William Howe. Howe was from Spencer, Massachusetts, and during the building of the Western Railroad he moved to Warren, where he lived within sight of the tracks. Later, he moved to Springfield. Howe received two important bridge patents in 1840 and 1846, respectively.

The basic Howe truss was a simpler and more durable structure than previous designs, and it also used less material to accomplish its goal. It shared the economic advantages of the Town and Long designs, but offered significantly greater strength, making it especially desirable for railroad applications. Some writers have considered the original Howe truss to be a transitional design because it combined wood and iron members. It evolved to become among the first truss designs adapted for all-iron construction.

Howe's patents described his bridge as a "truss-frame," which consisted of a system of diagonal timber braces under compression and held in place by vertical wrought-iron rods under tension, along with horizontal timber members under either tension (bottom) or compression (top). This design did away with two characteristics of Long's bridges: the diagonal supports below the main span that carried lateral forces from the lower chord to the abutments and the additional longitudinal king-post trusses above the upper chord at the central portion of the span.

Although Howe's initial patents were assigned in 1840, Howe had experimented with truss designs on a church in Warren, and he constructed his first

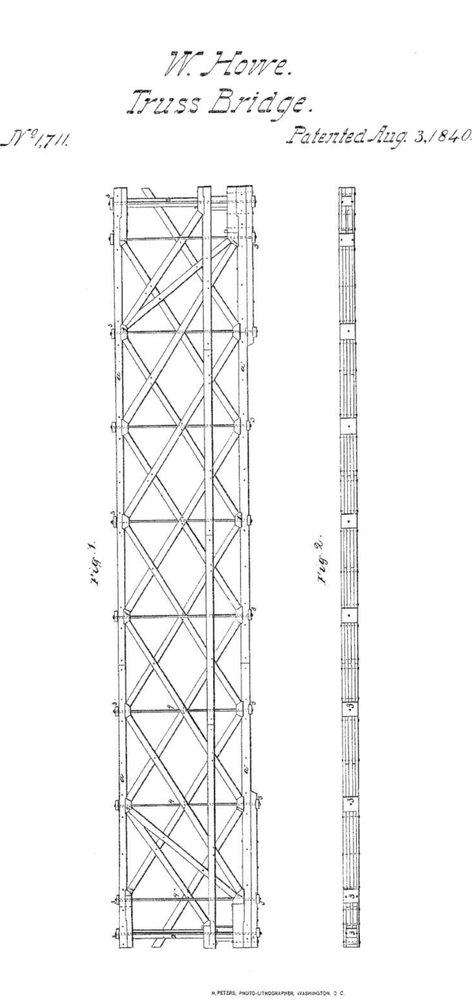

William Howe's 1840 patent was the basis for many railroad bridges built between the 1840s and 1860s. The first Howe truss bridge was a covered structure on the Western Railroad of Massachusetts, spanning the Quaboag River at Warren, just a short distance from Howe's residence. *U.S. Patent No. 1,711, August 3, 1840*

bridge on the Western Railroad between 1838 and 1839. The bridge was a covered design spanning the Quaboag River a short distance west of the village at a location later known as Sun Valley. Howe is believed to have supervised the construction of this bridge, and its success led to him designing a much larger bridge: Western Railroad's Connecticut River span at Springfield, Massachusetts, constructed between 1839 and 1840. This bridge was unusually large for the period and became one of the best-known early examples of the Howe truss. It provided the railroad with a relatively inexpensive solution to spanning the broad Connecticut River, which had represented one of the most serious obstacles for the line between Worcester and the Berkshires. Under Whistler's direction, Howe built this bridge with the help of his brother-in-law. Many authors cite Howe's colleague as Amasa Stone, but Plowden credits Daniel Stone as his principal partner on the Connecticut River Bridge. (Five Stone brothers were related to Howe through marriage and together acquired the rights to Howe's patents.)

Extracts from the "Report of the Western Railroad" published in the *American Railroad Journal and Mechanics' Magazine* in1843 describe bridges on the Worcester-to-Berkshires line and specifically highlight the Connecticut River crossing:

> *The Connecticut river [*sic*] bridge, and those westward of it, on the Western road, are of truss frames, of Howe's more recent patent, and they are constructed for one track only. The truss frames of all are covered in on the sides and top, and thoroughly white-washed. The entire flooring of the Connecticut river [*sic*] bridge is covered with tin, painted of a dark color. This bridge is 1264 feet long, of 7 spans, 180 feet each.*

In 1846 Howe received a patent for a stronger truss of the same basic design as his 1840 design, but employing large longitudinal timber arch beams that extended from the abutments at a level below the bottom truss chord to an apex where the arch beams were in contact with the top chords of the truss. Howe wrote in the patent "the arch beams which passes [*sic*] independently between the principle [*sic*] braces . . . of the truss frame and is sustained by the abutments in the usual manner, the principle [*sic*] brace being double, and at a suitable distance apart."

Howe's brother-in-law Amasa Stone formed an early bridge-building business that was exceptionally prolific, constructing a great many Howe trusses on American railroads during the mid-nineteenth century. Aiding Stone's success was his series of prominent administrative positions on a number of railroads, including the New Haven Railroad predecessor New Haven, Hartford & Springfield and a variety of lines incorporated under the umbrella of Cornelius Vanderbilt's New York Central empire. Among the lines Stone directed and built was the

Cleveland, Painesville & Ashtabula, which was melded into Vanderbilt's Lake Shore & Michigan Southern in 1873. The spectacular failure on December 29, 1876, of the first Howe truss that Stone adapted as a wrought-iron span caused a sensation that resulted in many changes in bridge building and design.

Among the notable flaws of early timber truss designs, in addition to the constant danger of steam locomotives starting fires, was susceptibility to destruction by ice floes during spring thaws and floods. Despite the move toward all iron, and later steel, for truss-bridge construction, some early timber trusses survived in railroad service into the twentieth century.

IRON TRUSS BRIDGES

In Britain railways were intertwined with the rapid growth of the iron industry. Not only did railways make commercial production of iron easier by vastly improving transport infrastructure, they also became one of the largest markets for iron. In eighteenth-century Britain, iron was developed as a building material; the world's first cast-iron bridge was erected in Britain by Abraham Darby working with John Wilkinson. Still standing, "Iron Bridge" was erected over the River Severn at Coalbrookdale, England, in 1779. It consists of five semicircular cast-iron ribs spanning approximately 100 feet.

The success of Iron Bridge led to many bridges of similar construction. Britain's famous bridge builder Thomas Telford was among the early proponents of cast-iron arch design. Although he did not build railway bridges himself, his works inspired similar structures used on early British railways.

Experience had taught bridge builders about the benefits and failings of cast iron versus wrought iron. Cast iron is well suited for distributing compressive forces and is resistant to corrosion, but is weaker under tension than wrought iron. Since an arch is supported by compressive forces, cast iron could be substituted for masonry. Like the masonry arch, the classic cast-iron arch rib was a product of empirical design and typically employed far more material than required to support the dead load of the bridge and the relatively light live loads of the period. However, because these early bridges were so heavily built, many have survived in service to the present day.

The first cast-iron arch built in the United States was a replacement bridge erected in 1836 on the National Road spanning Dunlap's Creek in Brownville, Pennsylvania. Yet, for a variety of reasons, cast-iron arches did not enjoy widespread application among U.S. railways. As domestic methods of production improved, however, iron became less expensive in America and thus a logical choice for bridge building. Wrought iron was the material of choice for iron bridge construction in America. While cast iron had been substituted for masonry in Britain, the properties of wrought iron made it well suited for tension members, and therefore it was substituted for timber in America. Wrought-iron tension members were used to strengthen timber trusses such as those in the Howe truss. In 1847, an adaptation of the Howe truss using a mixture of cast- and wrought-iron members was constructed in western Massachusetts on the North Adams Branch of the Western Railroad.

WHIPPLE, FINK, AND BOLLMAN TRUSSES

As railroad building gained momentum in America, there was an ever greater need for inexpensively built and easily constructed bridges of greater and greater strength. As locomotives and trains grew heavier and faster, engineers began to study the effects of the forces of trains crossing bridges. During this crucial period, bridge design gradually matured from an art to a science.

Bridge building had traditionally been an empirical art that relied upon trial-and-error, model building, and rules of thumb rather than mathematical analysis. With the development of sophisticated iron truss plans, quantitative mathematical research into bridge stresses emerged as a crucial factor in bridge design, and in this way, railroad applications began to drive the entire academic realm of civil engineering.

Squire Whipple, an engineer based in Utica, New York, is credited as the first to have published detailed analyses of bridge stresses. Whipple, like Howe and other pioneering truss designers, originally came from central Massachusetts. Henry Petroski sketches Whipple's life in his book *Engineers of Dreams* (1995). Born in 1804 in Hardwick, Massachusetts, Whipple

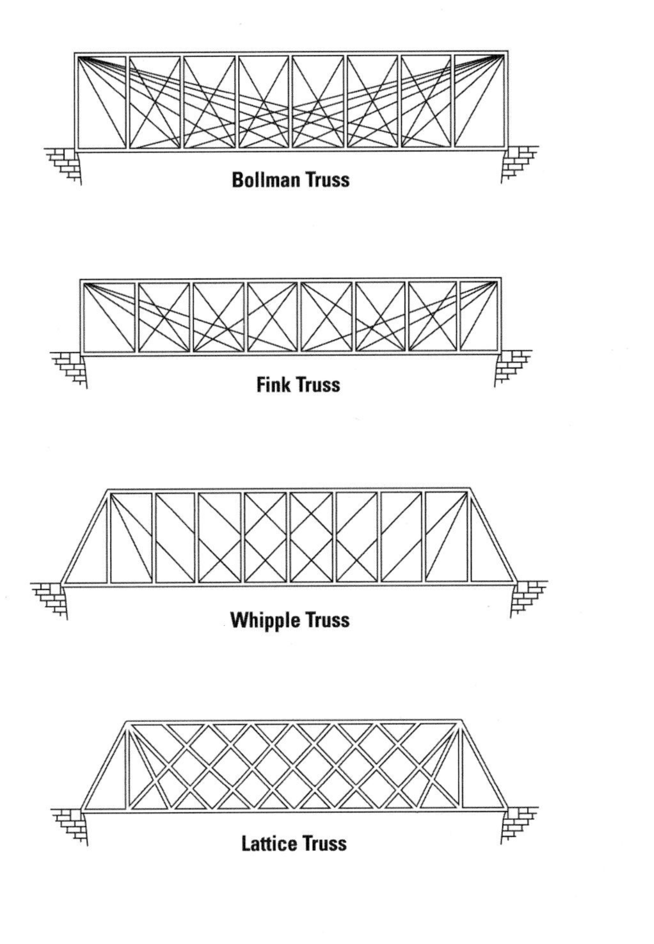

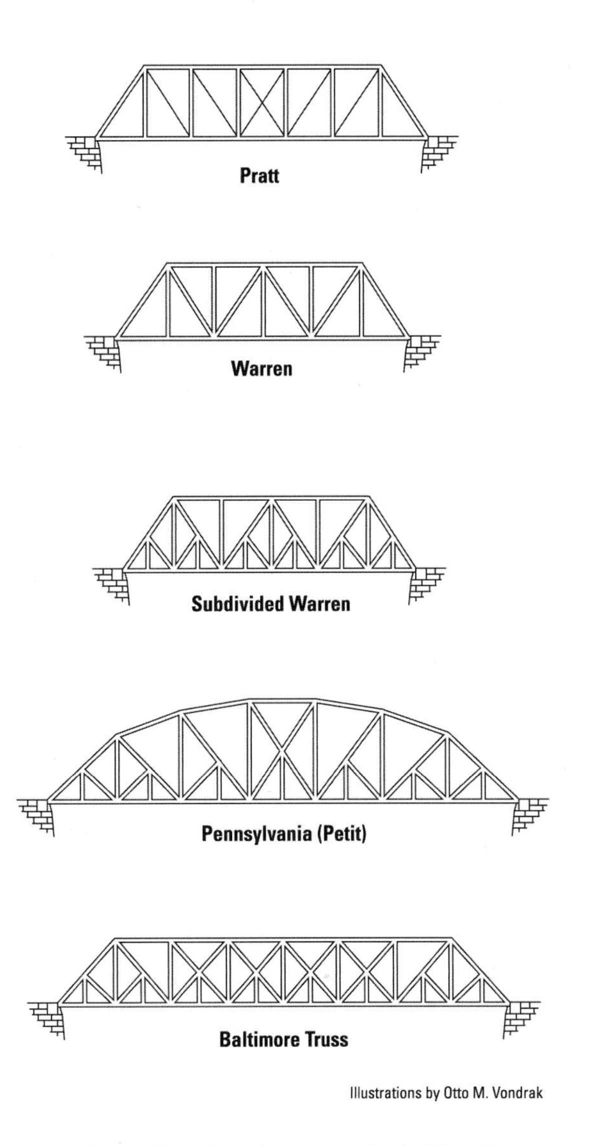

Illustrations by Otto M. Vondrak

was educated locally and at Fairfield, Connecticut, before attending Union College in Schenectady, New York.

Whipple was among the first to demonstrate and quantify the characteristics of wrought iron, showing that its resistance of tensile forces was not only significantly greater than that of wood, but four times greater than cast iron. In 1841, he received a patent for a bowstring iron truss. Whipple notes in United States Patent No. 2,064 that his invention, although called an "Iron-Truss," could be applied to wooden bridges as well. It consisted of an arched-iron top chord connected by vertical wrought-iron tie rods and posts to the bottom chord. In addition, were diagonal braces in X-like arrangements between the vertical members.

In 1846–1847 Whipple published "An Essay on Bridge Building," which detailed the stresses on the members of trusses. The book is considered a landmark in the history of bridge building and civil engineering, a groundbreaking work that led to scientific bridge building and established his reputation in the field. A host of detailed writings analyzing bridge design followed Whipple's initial publication. Among the better-known American writers was Herman Haupt, who had been writing on bridge design since the 1840s, and in 1855 published *The General Theory of Bridge Construction*. In 1873 Whipple published a much expanded book, *An Elementary and Practical Treatise on Bridge Building*.

Whipple's works were more than theoretical analyses. He was quick to apply his theories to actual bridge design. In 1847 he introduced a trapezoidal

An unusual application of the Whipple truss was this old span incorporated as a roof support above the snowshed-enclosed turntable on the Southern Pacific at Norden, California. Most of the snowshed complex used conventional wooden-beam and roof-truss construction measuring 16 feet wide and 22 feet tall, but the distance over the turntable required a greater span than the ordinary roofing arrangement could support. On December 9, 1990, an eastward SP freight rolls through the Norden sheds toward Tunnel 41. The sheds pictured here were demolished in summer 1993. *Brian Solomon*

iron truss design that was later adapted for railroad bridges. In later years, this bridge became known as the Whipple truss.

The classic Whipple truss features vertical tension members dividing each panel and uses rows of parallel diagonals facing each nearest end post, running from the top chord to the bottom chord across two panels. These are arranged to run to the center of the bridge on each side of the truss. In the most basic application, the pattern of diagonals is such that only at the center panel do tops of the diagonal members radiate out toward both end posts. However, in stronger applications of the design these diagonals can be continued across the length of the span. The arrangement of the diagonals forms a series of trapezoids. An additional set of diagonals at end posts cross run from the top chord to the bottom chord across just one panel, and so drop at a sharper angle. The combination of vertical tension members, with the diagonals radiating at two different angles from end posts, and the multiple-intersection diagonal arrangement help distinguish Whipple designs from both the Pratt and Warren designs that became standard in later years.

Similar to the classic Whipple truss is a multiple-intersection lattice-truss design that uses more

This former Chicago & North Western metal lattice truss at Oshkosh, Wisconsin, built in 1899, is typical of a type widely installed on the railroad in the late nineteenth and early twentieth centuries. The center truss is a movable swing span, and it is open in this September 2, 1995, photograph. *Brian Solomon*

diagonal members. These members slant in both directions across the full length of the truss, but without panels defined by vertical members. While the trapezoidal truss required further modification before it gained widespread acceptance as a railroad bridge a generation later, in the meantime, two other examples of iron trusses were pioneered on the Baltimore & Ohio.

B&O's early demand for bridges produced a generation of bridge designers and a variety of innovative designs. Whipple was briefly employed by B&O as a surveyor, giving him at the very least a passing connection with the railroad. B&O's eminent engineer, Benjamin H. Latrobe, helped train two engineers, Wendel Bollman and Albert Fink, both of whom emerged among the early generation of successful iron truss builders.

Bollman began working for the B&O as a young man and by 1848 had risen though the ranks to the position then described as "master of road." In 1850, he designed and built an original type of iron truss that spanned the Little Patuxent River at Laurel, Maryland. He was granted a patent for this design in January 1852. Although in his patent he describes his structure as a "suspension bridge," and while his design shares qualities with suspension bridges, it

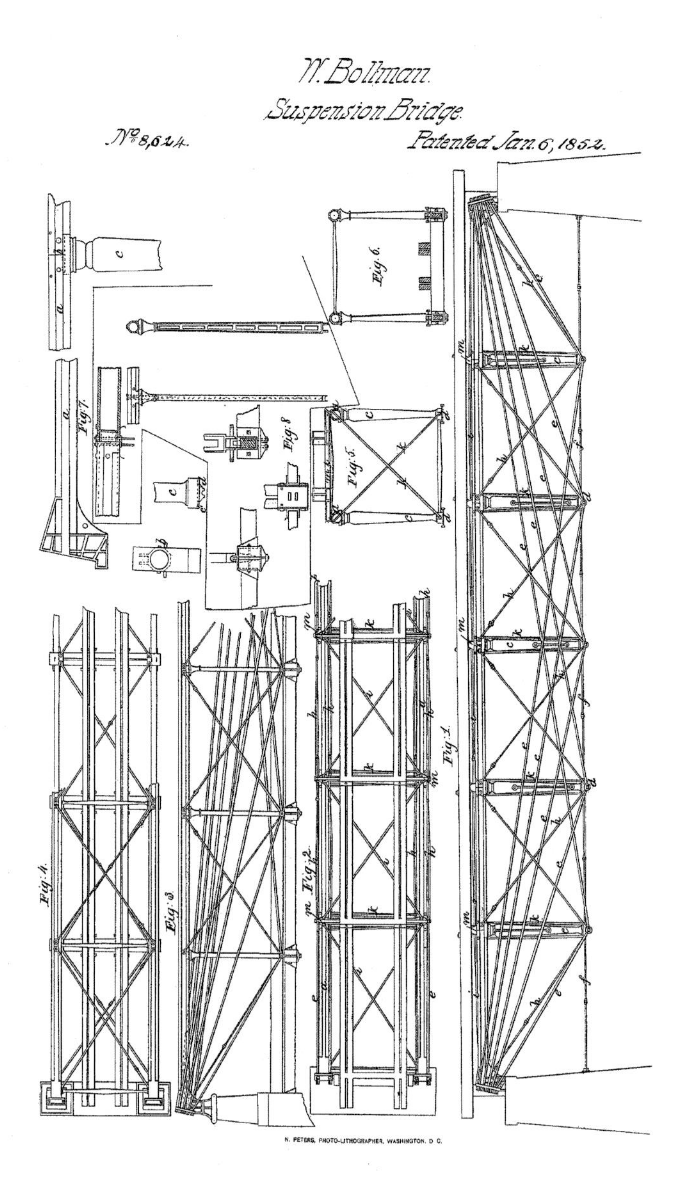

Many Bollman trusses were constructed on Baltimore & Ohio's lines between 1850 and the mid-1870s. In his patent, Wendel Bollman described his iron truss as a "suspension bridge." The patent contains an illustration of how Bollman's truss may be used as either a deck or a through truss. A deck bridge carries the railway on its top chords, with the bulk of the bridge located below rail level. *U.S. Patent 8,624, January 6, 1852*

A Guilford Rail System freight works across the bridge at Salmon Falls, New Hampshire, toward Portland, Maine, in March 2002. This unusual deck lattice truss (detail, below) has been reinforced for heavier axle loads with the addition of more rows of diagonal members. Despite the proliferation of more common standard designs, a number of unusual bridges from the steam era remain in service. *Brian Solomon*

The last surviving example of a Bollman-patent truss bridge (detail, right) is seen at Savage, Maryland, in the early hours of November 1995. Very few Bollman- and Fink-style iron trusses survived into the modern era. None remain in active service as railroad bridges today. *Brian Solomon*

is more closely related to iron truss bridges of the period. His patent provides an illustration of how his design may be used as either a deck truss or a through truss. The former type carries the railway on its top chords, above the superstructure of the bridge, with the bulk of the bridge located below rail level. The latter type has the rails riding on the bottom chords, and the bulk of the bridge above rail level, thus the train appears to travel "through" the bridge.

The Bollman truss used a combination of cast- and wrought-iron members with timber beam flooring for a cheaply built, easily assembled, and very strong structure well suited to contemporary railroad operations. Compression members, including the horizontal top chords, end posts, and intermediate vertical struts defining the panels, were all hollow, octagon-pattern cast-iron rods. The truss used a unique arrangement of wrought-iron tension rods that radiated from end posts on each side of the truss and ran longitudinally, connecting with the respective bottom chord member and intermediate vertical struts at each and every section point. Since these tension members radiated in

A tinted period postcard shows a single-span Bollman truss along the Potomac River near Harpers Ferry, West Virginia, on the Baltimore & Ohio Railroad. A more substantial multiple-span Bollman truss was used for the crossing of the Shenandoah River not far from this modest single span. *Author collection*

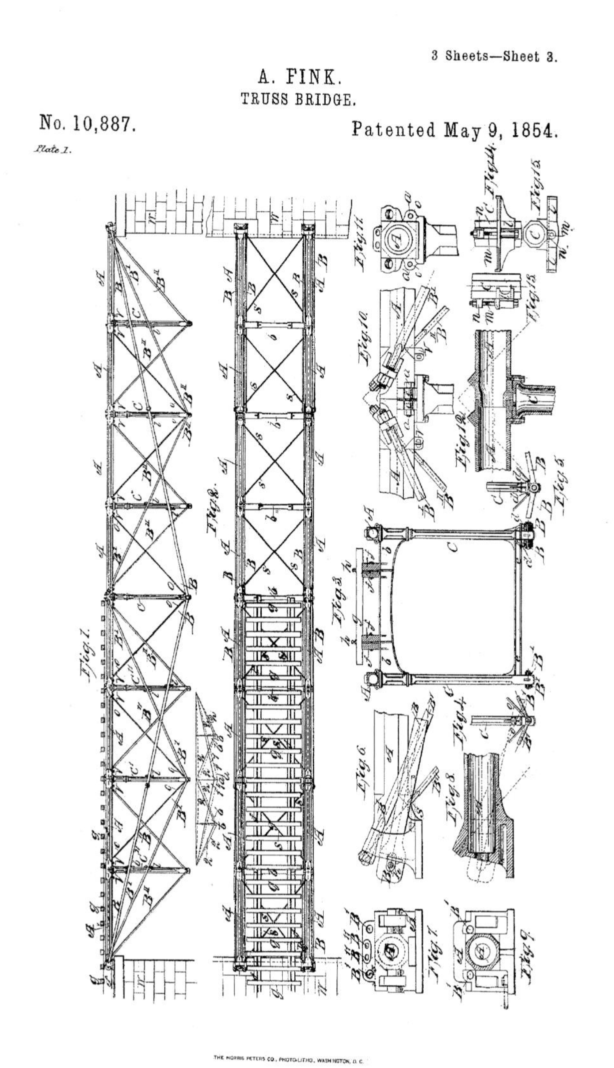

Albert Fink, like Wendel Bollman, was a Baltimore & Ohio engineer. Both men were protégés of B&O's Benjamin H. Latrobe, whose influence on their designs is undeniable. Fink's bridges, like Bollman's, used cast-iron rods for compression members, and wrought-iron tension rods. Many of Fink's bridges were built on B&O's western lines. After 1880, most Bollman and Fink trusses were replaced with heavier commercially designed iron or steel bridges, typically using Pratt and Warren trusses or plate-girder construction. *U.S. Patent 10,887, May 9, 1854*

both directions, the web-like tension rods made for a distinctive-looking bridge.

The improved factor of safety made possible by the redundancy of the design gave Bollman's design an advantage over other types of bridges that might suffer collapse in the event of a flawed member. The most prominent Bollman truss was the multiple-span bridge across the Potomac at Harpers Ferry, West Virginia, built in 1852. B&O adopted the type for many of its new spans on its eastern lines.

Albert Fink was born in Lauterbach (located in the modern-day Sachsen region of Germany) in 1827, the year B&O was chartered. He studied engineering and architecture in Darmstadt, graduating in 1848, then immigrated to Baltimore in 1849, where he soon joined B&O's engineering department. Among other work, Fink designed a truss bridge similar to Bollman's, for which he was granted a United States patent on May 9, 1854. Like Bollman, Fink employed hollow cast-iron rods for compression members and wrought-iron tension rods. Although the arrangement of members was distinctive and transmitted the stresses differently than Bollman's design, the kinship of these bridges is unmistakable. While the Bollman design was often built as both a through truss and a deck bridge, Fink's design was more commonly applied as a deck bridge. Among Fink's most famous trusses was a three-span structure across the Monongahela River at Fairmont, West Virginia. Each of these spans was reported to be 215 feet long.

After working for B&O, Fink went on to the Louisville & Nashville. Advances in truss design effectively rendered both the Bollman and Fink designs obsolete by about 1876, and very few of these bridges were constructed after that time, though many remained in service for decades.

The Bollman and Fink designs are both examples of double-intersection trusses, as was Whipple's trapezoidal truss, mentioned earlier. Although invented first, the Whipple truss didn't find favor as a railroad bridge until the 1860s, when the Lehigh Valley Railroad's chief engineer, John W. Murphy, designed a Whipple truss implementing pin-connected joints. Murphy is credited with pioneering wrought-iron pin-connected joints on a bridge at Phillipsburg, New Jersey, in 1858. This innovation had a twofold advantage: (1) pin connection made erection of the truss much easier because primary members could be manufactured in factories and easily assembled on site, and (2) the connections insulated bending stresses. The pin-connected Whipple trapezoidal truss was a winning combination that rapidly emerged as one of the most popular bridge designs in the latter nineteenth century. The type was further adapted to good advantage by renowned bridge builder Jacob H. Linville.

Unlike the heavily constructed metal trusses commonly used today, mid-nineteenth-century truss bridges were designed for the lighter axle loads of the period and display a much airier style of construction. Photographed in 1992, this pin-connect Whipple trapezoidal truss over the Connecticut River served a branch of the New Haven Railroad running to Turners Falls, Massachusetts. Notice the relatively thin tension members and pin connections at the deck level. *Brian Solomon*

Below: The truss in the photographs on this page is one of the few surviving Whipple trapezoidal types. It is part of a three-span bridge over the Connecticut River at East Deerfield, Massachusetts. In later years this branch was operated by Boston & Maine. Tracks were lifted years ago after regular freight operations ended in the 1970s. Although the bridge sat disused for more than two decades, at the time of this March 4, 2007, photo, it was being converted for use as a hiking trail. *Brian Solomon*

Above: The Davenport, Rock Island & North Western's Mississippi River Bridge at Rock Island, Illinois, was designed by C. F. Loweth and built by the Phoenix Bridge Company of Phoenixville, Pennsylvania, in 1898. Originally a joint property of Burlington and Milwaukee Road, DRI&NW was later jointly owned by Burlington Northern and Canadian Pacific's Soo Line subsidiary. In the mid-1990s, CP divided DRI&NW trackage with BNSF and absorbed its share of the line. A few years later, CP spun off its Chicago–Kansas City route, including its remaining DRI&NW trackage. *Brian Solomon*

Opposite: An unusual arrangement of Pratt through trusses supports the DRI&NW bridge on a curve. A swing span at the center allows water traffic to pass. On June 11, 2004, a former Soo Line GP9 leads an Iowa, Chicago & Eastern freight eastward across the bridge. *Brian Solomon*

PRATT AND WARREN TRUSS

In the latter decades of the nineteenth century, in general, truss designs were designed, built, and installed by commercial bridge companies. By this time, many of the early experimental truss varieties were no longer deemed appropriate, so the majority of commercially built trusses used the Pratt and Warren patterns, or variations thereof.

The Pratt truss was among the several American wooden truss patterns developed in the 1840s. Named for its designers, Thomas W. Pratt and Caleb Pratt from Norwich, Connecticut, its primary diagonal members face toward the nearest end post and away from the center of the bridge, and panels are divided by vertical members. As with the Whipple truss, diagonal braces are in tension. The end posts are diagonal and face toward the center.

The Pratt truss proved well suited for iron and, later, steel members. Later variations include the Parker truss, which has an inclined top chord in an arch-like pattern, but fabricated from straight beams; the Baltimore truss that features added half-length diagonal subdivisions; and the Pennsylvania truss, which, in addition to the added half-length diagonals, uses inclined arch-like top chords.

Continued on page 48

1898
D.R.I.& N.W.
RY.
H.B.SCHULER
PREST.
F.P.BLAIR
VICE PREST.
120
120

315
Boston&Maine

Above: A rare surviving example of a pin-connect Whipple trapezoidal deck truss at Easton, Pennsylvania. The Easton & Northern Railroad Company used this now-abandoned line to reach the Lehigh Valley main line. *Brian Solomon*

Left: Boston & Maine crosses the Hudson River at Stillwater near Mechanicville, New York, on this rather substantial double-track Warren deck truss resting on reinforced concrete piers. On November 4, 1982, B&M GP40-2 No. 315 works as a helper shoving on the caboose of symbol freight MELA (Mechanicville, New York, to Lawrence, Massachusetts). Many railroads continue to use bridges designed and built in the steam era. The comparative cost of replacing such structures is much higher today than 80 to 100 years ago. *Jim Shaughnessy*

Erected in 1902, an abandoned Pratt through truss over the former New York Central main line west of Newark, New York, uses a skewed arrangement. Compare the sharp angle of the lateral connection between the end posts with the other lateral struts. This bridge had been used to carry a light branch to Marion, New York, and was subsequently removed to improve clearances on the Water Level Route below. It was photographed here on November 1, 1986, years after the line was closed. *Brian Solomon*

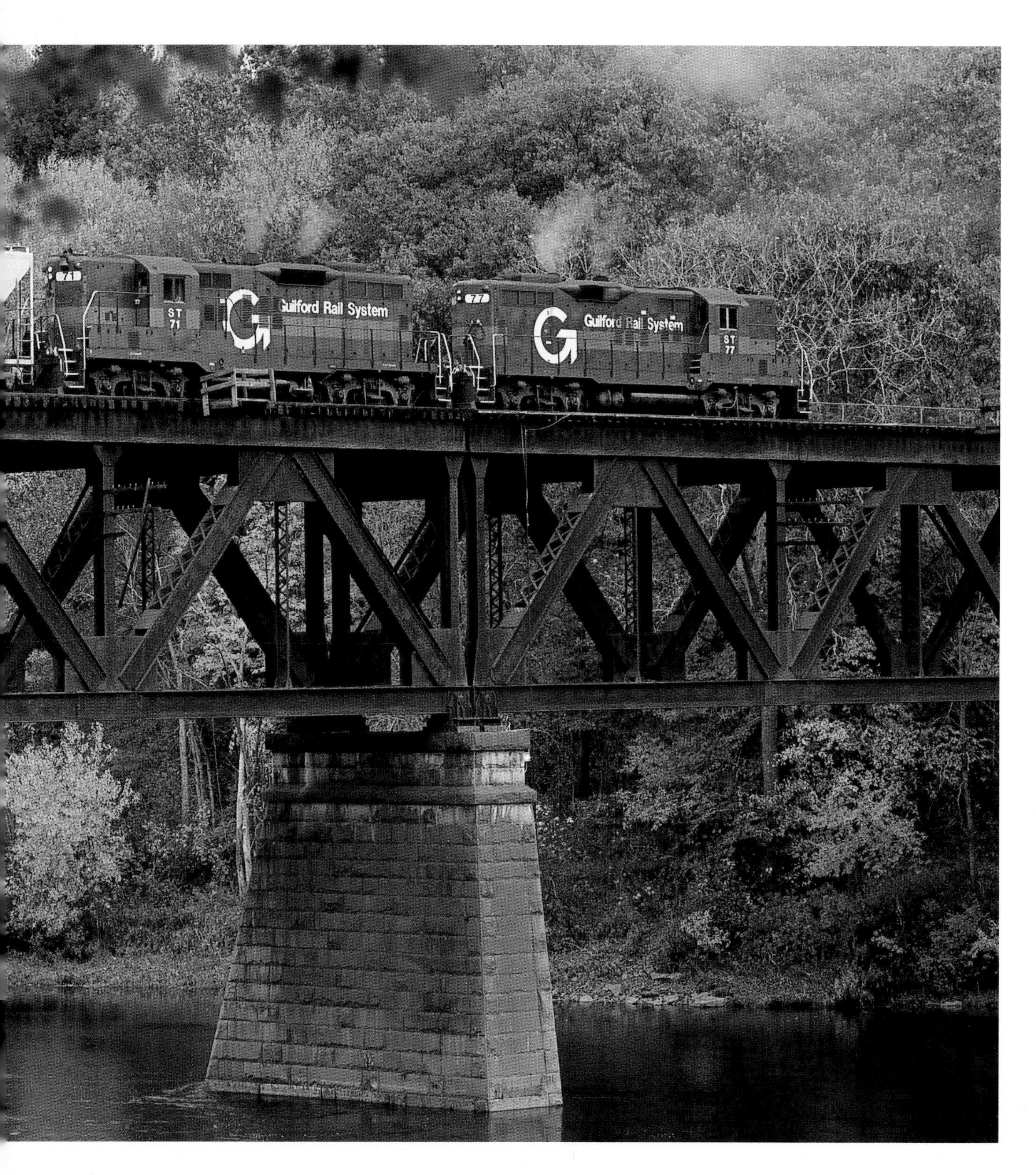

Boston & Maine's Connecticut River Bridge at East Deerfield, Massachusetts, consists of a heavy, double-track, multiple-span Warren deck truss on solidly built masonry piers–typical of late-era bridges on the line. On October 18, 2004, in a routine move that occurs several times daily, the East Deerfield Yard hump engines pull a cut of cars out of the yard and across the bridge. *Brian Solomon*

In the late afternoon on December 14, 1992, Green Mountain Railroad's Bellows Falls–to–Rutland, Vermont, symbol freight XR-1 gingerly crosses the Warren deck truss at Cuttingsville, Vermont. This is one of many bridges on the line dating back to the days of the Rutland Railroad. *Brian Solomon*

Continued from page 42

James Warren and Willoughby Monsoni developed the Warren truss in Britain in the late 1840s. The design requires an even number of panels. The diagonals alternate facing one direction and then the other, creating a pattern of equilateral triangles. Typically, end posts are inclined, representing one side of an equilateral triangle. Variations of the Warren pattern use vertical members to divide the panels; these add strength. Another type is the double-intersection Warren, which uses a second set of diagonals in a lattice arrangement for greater strength.

IMPROVED TRUSS CONSTRUCTION

As railroads demanded ever-stronger bridges, trusses evolved to accommodate greater loads with less routine maintenance. Wrought-iron designs fell out of favor and all-steel construction became standard. For many years, pin-connected construction was favored because of ease of assembly. However, the system had structural disadvantages. In a pin-connected truss, the pins connect the various members at joints. These members, including both tension and compression members, have holes in them to accept the pins. A pin itself is a "shear" member, meaning it resists opposite forces in a straight line across the diameter of the pin itself. This presents several problems. Because they are keys to the structure of the bridge, pins must be inspected, a difficult job that typically requires jacking up the truss, which relieves stresses and allows the pins to be removed for inspection. This operation generally cannot be accomplished with the bridge in service, and thus results in unacceptable delays. More serious is that if a pin fails at the wrong moment it can result in the catastrophic failure of the whole structure.

In a visual and technological contrast, Amtrak's General Electric–built Genesis diesels lead the *American Orient Express* luxury excursion train across Vermont Rail System's Cuttingsville Trestle on October 5, 2004. The diesels use state-of-the-art monocoque body-shell design, while the bridge is a nineteenth-century adaptation of the Warren-style truss. In both cases, the root technology was imported from Europe and adapted for American purposes.
Brian Solomon

The first railroad bridge across the Connecticut River at Springfield, Massachusetts, was the second Howe truss bridge ever built, an exceptionally long covered wooden truss. The present double-track bridge consists of parallel seven-span, single-track steel trusses built on common masonry piers. On the morning of December 2, 1992, the river was unusually placid, allowing a reflection of the bridge and the Springfield skyline. *Brian Solomon*

Built in 1903, the atypical three-span truss over the flood-prone Sinnemahoning Creek at Keating, Pennsylvania, is on a skew and features vertical end posts. Originally, a long, single-track covered timber truss was located here, roughly 160 miles from Harrisburg, likely on the same masonry piers. This view, looking toward Renovo and Harrisburg on the former Pennsylvania Railroad at the west end of the truss, was taken on October 11, 2003. *Brian Solomon*

Detail of the portal strut, top chords, and lateral braces on the former Pennsylvania Railroad skewed truss bridge over Sinnemahoning Creek at Keating, Pennsylvania. *Brian Solomon*

After 1900 railroads moved toward riveted connections for major joints. These use steel plates known as gusset plates that cover joined members on one or both faces and are riveted to all respective members. Rivet-connected trusses became standard after World War I. Many of the more heavily built riveted trusses used the Warren design, usually with center vertical hangers.

The introduction of high-strength bolts in place of rivets was introduced in 1960. Rivet replacement is labor intensive work, and as the price of labor increased, the need to lower labor costs became a more important part of bridge construction and maintenance. While bolts are more expensive than rivets to manufacture, they are much cheaper to install, adjust, and replace. Bolts do not require any special tools and can be easily worked on site.

An early example of a bolted bridge was Burlington's 2,500-foot-long, 14-span replacement crossing of the Mississippi River at Quincy, Illinois, installed in 1960 and featured in the November 28, 1960, issue of *Railway Age*, which wrote that it consisted of six deck-girder spans, seven Warren-truss deck spans, and one through Warren span.

"The final two-thirds of the bridge was assembled with high-strength bolts instead of rivets," the magazine reported. "The switch came about this way: When about one-third of the bridge had been erected, with rivets used for the connections between members, it became apparent that a lack of experienced riveting crews would prevent the structure from being completed on schedule."

Today, on many railroads, when a riveted truss is rebuilt, old rivets are replaced with bolts. A bolted truss can often be identified by the hexagon bolt heads visible on the gusset plates and at various connections instead of rounded rivet heads. However, this is not a foolproof method of spotting a riveted truss: Many structures covered by historic preservation legislation must retain their visual characteristics following rebuilding, and such trusses use specially designed and installed round-head bolts to maintain their appearance.

This former Central Railroad of New Jersey bridge at Easton, Pennsylvania, is an example of a pin-connect Pratt through truss built on a skewed alignment. *Brian Solomon*

A detailed view of the pin-connected eyebars on the former Central Railroad of New Jersey Pratt through truss at Easton, Pennsylvania, pictured above. *Brian Solomon*

On a frosty February 2005 morning, the rising sun illuminates Canadian National's double-track Pratt deck truss spans Riviere Richelieu at Beloeil, Quebec, as the fog burns away. Connections are riveted together with gusset plates. With the Pratt truss, primary diagonal members are angled in the same direction on each side of the span center; with the Warren truss, primary diagonals alternate direction the length of the span. *Tim Doherty*

A Guilford Rail System local from Rumford, Maine, rolls across a modest Pratt through truss on Maine Central's Rumford Branch over the Dead River near Leeds, Maine, on April 5, 1997. This is an example of riveted-joint construction, used in place of older pin-connected designs. *Brian Solomon*

In the shadow of the massive cantilevers for Interstate 95 is the former New Haven Railroad Thames River Bridge between New London and Groton, Connecticut. Constructed as a replacement for an earlier bridge, work began during World War I and was finally completed in 1920. Of the five spans, four are Pratt through trusses using inclined top chords, while the middle span is a Strauss heel trunnion bascule using a Warren truss. On Halloween Day 1997, a pair of Amtrak F40PHs leads a Boston-bound train near the east end of the bridge at Groton. In 2007, this historic bridge was in the process of replacement, with a new vertical lift bridge being built to replace the bascule. Amtrak, which owns and operates the line, expected to have the new bridge in operation by mid-2008. *Brian Solomon*

This view made in 1986, shows the Groton end of the former New Haven bridge over the Thames River seen above. This double-track bridge, built in 1920, uses a substructure designed for a four-track line. *Railway Age* noted at the time of construction that New Haven anticipated the need for greater capacity. Unfortunately for the railroad, that capacity was never required, due to the rapid rise of highway transportation–as evident by the dual cantilever bridges at right that now carry Interstate 95. This riveted Pratt through truss span is 330 feet long and uses inclined top chords for greater strength. At mean low tide it provides a 33-foot, 4-inch clearance. *Brian Solomon*

CHAPTER 3

Trestles and Metal Viaducts

The trestle was well known to early American railroad builders, predating the railroad. David Plowden states that the earliest known pile trestle in America was built in 1761 by Major Samuel Sewell at York, Maine.

WOODEN PILE TRESTLES

Pile trestles have been popular with railroads because of their simple design, low cost of construction, and ability to be erected quickly from readily available materials without the need for complicated tools. Wooden pile trestles range from spans just long enough to cross small streams to miles-long crossings of vast bodies of water. One of the most extensive, albeit wholly unsuccessful, applications of pile trestle–work was by the Erie Railroad on its Susquehanna Division. Erie's first president, Eleazar Lord, thought there would be substantive cost savings in substituting wooden pile

Left: An Illinois & Midland freight eases across a timber pile trestle with a ballast deck near Oakford, Illinois, in May 2006. This trestle crosses the Sangamon River flood plain. The timber pile trestle is one of the most straightforward varieties of bridge construction and has remained an effective solution in situations where foundations are cost prohibitive. The piles are in compression and carry the load to the ground. Lateral bracing is primarily for stability. A relatively low pile trestle such as this one often does not require longitudinal bracing, which is used to prevent swaying on taller trestles. *Steve Smedley*

Long Island Rail Road electric multiple-units roll across the Long Beach wooden pile trestle at Island Park, Long Island, on August 2, 1959. Wooden pile construction is often used across tidal waterways where the cost of laying foundations is impractical. Notice the high-voltage electric third-rail along the outside of the crossties. *Richard Jay Solomon*

On October 29, 1994, a Southern Pacific crew is hard at work repairing the bridge over the San Jacinto River near Crosby, Texas, on the Sunset Route following a major washout that destroyed more than 500 feet of the 2,000-foot crossing. The bridge comprised mixed construction, using timber and steel piles, along with reinforced concrete. The railroad used the opportunity presented by the washout to build a more modern bridge using pre-stressed concrete-girder spans. Here, a new 90-foot steel pile is positioned for the pile driver. Later, the 24-inch-diameter pile will be filled with concrete for added strength. *Tom Kline*

A bridge and building crew on Illinois Central Gulf repairs a pile trestle at Munger—between Bartlett and South Elgin, Illinois—in April 1976. The relative ease of construction and low cost of material has made wooden trestles a popular choice for railroad bridges. *Steve Smedley*

This October 1985 photograph depicts a timber pile trestle on Burlington Northern's Houston Subdivision spanning Cypress Creek south of Louetta, Texas. The trestle uses a typical format: Each "bent" consists of five piles, with outside posts battered (angled) toward the center, and the center post vertical. The lateral braces, known as sway braces, are arranged in an X pattern on each bent, with one diagonal member bolted to each side of the bent in an alternating fashion. The sway braces are arranged in multiple tiers. On top of the piles are lateral stringers, known as caps, which support rows of longitudinal stringers that make up the flooring. The ties are attached to the flooring, and timber guard rails are placed beyond the running rails. In early 2007, plans were underway to replace this timber structure with a modern prestressed concrete bridge. *Tom Kline*

trestle–work in place of earthen embankments and conventional grading. An estimated 100 miles of pile construction was ultimately completed. The concept, although novel, was fundamentally flawed and not an effective method of construction. The wasted time and expenditure cost the Erie dearly in its formative years, and the line had to be largely rebuilt using conventional construction with cuts, fills, culverts, and bridges.

By the late nineteenth century, major bridge construction using iron, steel, and concrete was largely the domain of specialized commercial bridge-building firms that contracted with railroads. Construction of timber trestles, on the other hand, often remained the domain of the railroads' bridge and building departments, since engineering and building timber trestle structures is comparatively simple and straightforward. Although not as common as they once were, timber trestles of various sizes survive to this day on both lightly traveled branches and main lines.

Basic wooden trestle construction consists of the substructure and the deck. The substructure commonly uses either pile (timbers driven into the ground to develop strength) or frame construction, and sometimes both. The deck may be either open or ballasted; an open deck is cheaper to construct and allows for a lighter dead load, but is not as durable or as easy to maintain as a ballasted deck. Treated timber, typically using creosote, minimizes deterioration of members from rot. Pile trestles offer an advantage in situations where it isn't practical or cost-effective to build solid foundations, such as across shallow

Amtrak No. 11, the *Coast Starlight*, rolls across a double-track pile trestle over an inlet along Puget Sound at Steilacoom, Washington, behind a pair of SDP40Fs in March 1978. This heavily braced wooden trestle on Burlington Northern's former Great Northern line was adequate for the demands of traffic at the time, but has since been replaced with a concrete bridge. *Tom Carver*

bodies of water, including swamps, tidal estuaries, and lake beds.

Specific plans for pile construction vary depending on location requirements and the preferences of the railroad building it. In *Railroad Engineering* (1912), Walter Loring Webb provides detailed information on design and construction of wooden trestles in the period 1908–1912. Webb explains that the practical limit for wood pile construction is about 30 feet above ground; for taller bridges a wooden frame trestle is more practical. He notes that "usually four piles are considered sufficient for single track, although more are sometimes used. The inner piles are always made vertical but the outer piles are sometimes battered [angled inward] so as to give the trestle greater resistance against lateral thrust." Higher trestles and those built to sustain greater forces—either as the result of heavier trains or greater speed—require more substantial bracing than piles for simple culverts.

Today, wooden, concrete, and steel piles continue to be of paramount importance in railroad construction and can be driven by diesel-hydraulic hammers or vibrated into the ground using specialized equipment.

WOODEN FRAME TRESTLES

In most respects, frame trestles are similar to pile trestles, except, where the piles are driven directly into the ground, the frames rest on foundations, which may be of stone, concrete, wooden cribs, or indeed, wooden piles. Also, frame trestles require lateral and longitudinal bracing to maintain their structure regardless of their height.

A northward Florida East Coast freight crosses the pile trestle at Stuart, Florida, on April 22, 1978. The bridge uses steel-girder pilings encased in concrete to support a plate-girder deck. A basic trunnion-style plate-girder bascule span allows for the passage of pleasure boats from the Intercoastal Waterway into the Atlantic Ocean. *Brian Jennison*

At 12 miles, Southern Pacific's pile trestle across the Great Salt Lake was considered America's longest railroad bridge. In this view, a passenger train departs Midlake, the location of a train-order station and passing sidings. The trestle was abandoned when a multimillion-dollar fill was constructed parallel to it in the 1950s. *Voyageur Press collection*

Buffalo & Pittsburgh freight, symbol BT3 led by three vintage GP9/GP18s works across the Iron City Trestle at Chicora, Pennsylvania, at 4:55 p.m. on September 6, 2005. This basic type of multiple-tier, wooden-frame structure, using four timber piles per bent and an open deck, has served American railroads for more than 150 years. Despite more-advanced bridge designs, the classic frame trestle survives on lightly traveled lines where more-expensive construction is not justifiable. *Patrick Yough*

Northern Pacific Bridge No. 43 was a multiple-tier, S-shaped wooden frame trestle on the Coeur d'Alene Branch between St. Regis, Montana, and Wallace, Idaho. In this circa 1905–1910 view, a modest-sized locomotive, likely a Ten-Wheeler, leads an eastward passenger train. NP rebuilt the bridge twice: in 1903, when it was partly destroyed by an avalanche, and in 1910, when it was consumed by a forest fire. The bridge required 600,000 feet of timber. *Author collection*

The San Manual Arizona hauls copper ore concentrates 29.4 miles from the mine and smelter at San Manual to Hayden. The connection at Hayden used to be with Southern Pacific, but was later spun off to the Copper Basin Railway, which now takes traffic to the former SP (now Union Pacific) at Magma Junction, southeast of Phoenix. On April 8, 1995, San Manual Arizona's GP38s growl across the timber frame trestle at San Manual, near Mammoth. *Brian Jennison*

Among the most impressive timber frame bridges to survive into the modern era was the curved Half Moon Trestle on the Camas Prairie in Idaho. At the time of this 1994 photo Camas Prairie was jointly operated by BN and UP. Unfortunately, this line, which featured several large timber trestles, is no longer in service. Here, the westbound Grangeville Turn returns to Lewiston from Grangeville, Idaho, dropping downgrade into Lapwai Canyon. *Tom Carver*

Loaded coal hoppers roll across the Indiana Railroad's Tulip Trestle, located in Greene County, Indiana. American Bridge Company built the 2,306-foot, tower-bent-supported, plate-girder steel viaduct for the Illinois Central. It opened for service 1906. Eighty years later, Illinois Central Gulf spun off the line to Indiana Railroad. At its highest point, the viaduct is 161 feet above Richland Creek. *Steve Smedley*

A primary advantage of frame trestles is that they can be constructed in multiple tiers and so used for much taller bridges. The taller the trestle, the more bracing is required. Joints between timbers may use the mortise-and-tenon system (where timbers are cut and fitted to another) or more-substantial joiners such as heavy wooden dowels, iron or steel bolts, or iron or steel plates.

While frame trestles have been used all over the country, they are probably best known for their applications in the West, where precariously tall frame bridges were built as temporary structures when lines were hastily pushed across the frontier and rugged mountain terrain.

At 110th Street, near the northwestern corner of Central Park, stood "Suicide Curve," an imposing edifice of Manhattan's nineteenth-century rapid transit viaducts. Completed by Manhattan Railway Company in 1879, its rails were among the loftiest ever for transit at that time—an elongated S curve more than 100 feet high between 9th and 8th avenues. The El tracks connected directly with the New York & Putnam Railroad, later New York Central System, at the old Polo Grounds (erstwhile home of the former New York Giants baseball team). Drip pans protected pedestrians below from grease and oil. *Richard Jay Solomon collection*

Rapid Transit Viaducts

By Richard Jay Solomon

The most extensive networks of iron and steel viaducts have been built for elevated rapid transit systems in just a handful of U.S. cities, principally New York, Chicago, Boston, and Philadelphia. Including electric interurban and city trolley lines, the *1912 Street and Electric Railway Census* tabulated over 400 track miles on metal elevated transit viaducts in the United States. The peak for such viaducts was reached in the 1920s, before many rapid transit "els" were razed and trolley systems were abandoned.

In 1867–1868, the West Side and Yonkers Patent Railway became the first company to build an iron rapid transit viaduct in New York City. This pioneer line consisted of a single track supported by wrought-iron beams resting on single, rather flimsy iron poles set into the sidewalk. Originally cable-operated, the route served as a prototype for nineteenth century elevated railways. Eventually, it became Manhattan Railways' 9th Avenue El line, was much rebuilt and enlarged in succeeding decades, and finally was razed in 1942 as scrap for World War II munitions.

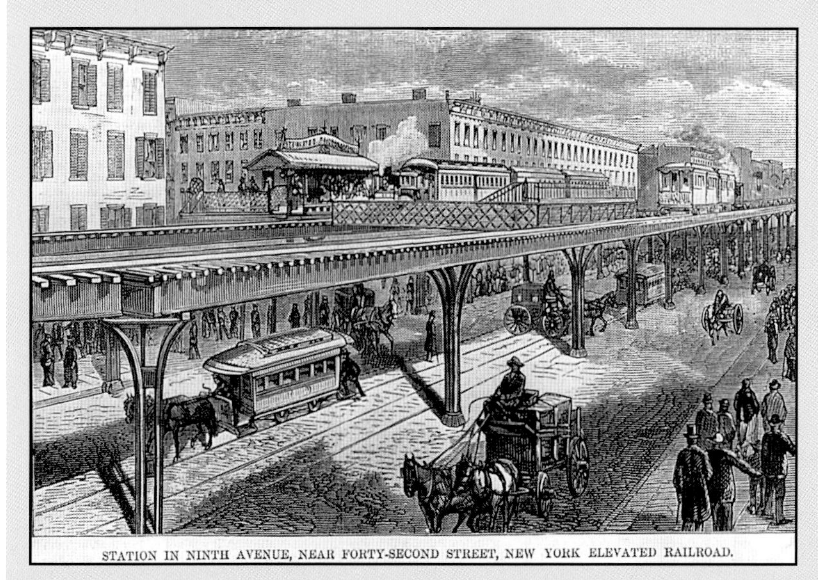

STATION IN NINTH AVENUE, NEAR FORTY-SECOND STREET, NEW YORK ELEVATED RAILROAD.

In 1868–1869, the world's first elevated rapid transit viaduct was built in New York City: a single-track, cable-operated line over the sidewalk on Greenwich Street, later extended up 9th Avenue. Supported on relatively thin wrought-iron girders, it was later completely replaced, rebuilt by the New York Elevated Railway Company as a double-track structure, and steam-operated. It is shown at 42nd Street in this woodcut from the late 1870s. In 1913–1918, a third express track was added and the viaduct was reinforced with more-modern steelwork. The El ceased operation in 1940, razed for munitions during World War II. *Richard Jay Solomon collection*

The next generation of elevated viaducts in Manhattan and Brooklyn, built between the 1870s and 1890s, was more substantial. While earlier designs consisted of wrought-iron lattice trusses supported on lattice vertical bents, later structures were made of steel. In the late nineteenth century, some 45 route miles of such el structures were built in New York and Brooklyn (separate cities until 1898) by several private companies. These els were operated with specially designed steam locomotives until the period 1900–1903, when electrification was implemented.

While the early iron viaducts were designed for lightweight vehicles (and locomotives), standard railroad coaches were not much heavier than el cars at the time. It is notable that, for a short while, the Long Island Rail Road ran steam trains over Brooklyn's 5th Avenue El to the Brooklyn Bridge approach and electric multiple-units via the Broadway–Brooklyn El over the Williamsburg Bridge to Manhattan. Similarly, the New York & Putnam (later New York Central) ran through coaches originating as far away as Boston, via The Bronx, and thence down the 9th Avenue El to the Battery. The Manhattan els even carried Wells Cargo freight in the early years of the twentieth century.

In Chicago, the Loop "L" viaduct carried interurban electric trains for the Chicago, North Shore & Milwaukee from as far as Milwaukee—a route that lasted until 1963.

By the 1890s, advances in steel fabrication made plate-girder viaducts feasible for rapid transit, a design that was pioneered for Chicago's South

Built by the Gilbert Elevated Railway Company in 1876 with ironwork by the Edge Moor Iron Company of Wilmington, Delaware, New York's 6th Avenue El was the first transit viaduct to use a patented truss design, somewhat more substantial than rival New York Elevated's earlier 9th Avenue structure. The photo looks north at 9th Street in Greenwich Village just before operation ceased in 1938 and construction began on the replacement IND 6th Avenue Subway.
Richard Jay Solomon collection

Side L and for later portions of the Manhattan el's "suburban" extension into the Bronx. More-modern viaduct construction, using mostly prefabricated and modular plate-girder steel components, became an essential element for the vast expansion of New York's rapid transit system from 1905 to the 1920s. An additional 50–60 route miles of steel viaducts were built as extensions to the original 1904 Interborough Rapid Transit Company (IRT) subway and for the 1913 "Dual Contracts" between New York City, the IRT, and the Brooklyn Rapid Transit Company (later Brooklyn-Manhattan Transit, or BMT). (Depending on how one defines terms, about half the "subway" system at the time was not strictly underground.)

Coincident with that rapid transit expansion, several of the original lattice-truss iron els in Manhattan and Brooklyn were strengthened, triple-tracked, and connected with the new subway lines. Virtually all of the earlier nineteenth-century els in New York City have been razed, most replaced by the Independent System (IND Division) subway expansion in the 1930s, though a few segments of original lattice iron viaducts are extant. Most of the Dual Contract viaducts are still in use today, with some extraordinarily complex webs of steel to behold, a prime example being the junction at East New York.

Reinforced concrete viaducts offered noise control, permitting ballasted roadbeds (concrete deadens the vibrations better than steel). However, concrete platforms added about 75 percent to the cost of steel viaduct structures, according to J. A. Waddell's 1916 tome, *Bridge Engineering*. Though the author admits that while "elevated railroads are an offense to the eye, an annoyance to the ear, and a general nuisance in every sense," and ballasted track on a "slab of reinforced concrete" would at least suppress noise, because of cost, concrete construction was not commonly used for early rapid transit viaducts. Philadelphia's Market–Frankford El and the now-razed Green Line light rail El in Boston were the few such metal structures built during that era with ballasted platforms.

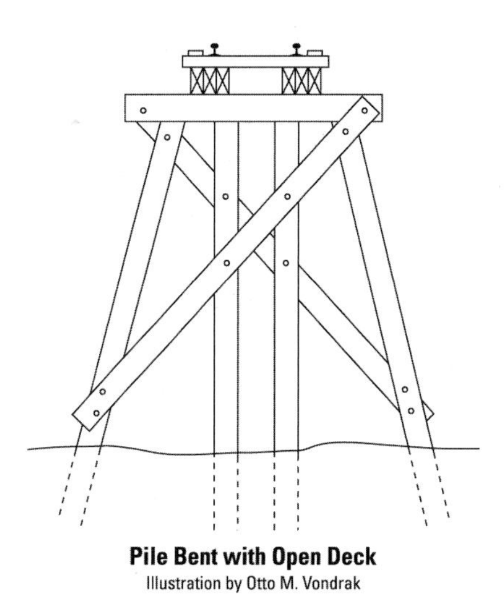

Pile Bent with Open Deck
Illustration by Otto M. Vondrak

IRON AND STEEL VIADUCTS

There are no universally accepted definitions classifying a trestle or distinguishing a trestle from a viaduct. For this book, a trestle may be considered as a continuously supported bridge for which the supports are relatively evenly spaced but not grouped. In general, a viaduct may be viewed as any multispan bridge. The iron and steel structures described below are often referred to as "viaducts," distinguishing them from single-span trusses, suspension bridges, and metal arches, although often these viaducts were built in conjunction with other types of bridge construction. Further, regional and railroad-specific terminology has varied. In some instances, railroads called steel viaducts "trestles," such as the cases with Somerset Railroad's Gulf Stream Trestle and Illinois Central's Tulip Trestle, both of which were essentially standard tower-supported plate-girder viaducts. Others have referred to these types of viaducts as "bridges" without further qualification, an example being Chicago & North Western's Kate Shelley Bridge. Further confusion results when some lines refer to perfectly ordinary multispan deck-truss bridges as "trestles."

The lineage of the metal viaduct has been clouded by time, lacking the clarity of origin and continuity of design found for the truss bridges described in Chapter 2. Compared with specific truss designs, the literature on metal viaducts is sparse. This is unfortunate because this bridge type played an important role in American railroad construction and often has been the subject of photographs. In *Bridges: The Spans of North America* (2002), David Plowden does an admirable job establishing an early chronology for applications of metal viaducts. He traces the earliest forerunners to bridges on Baltimore & Ohio's Cheat River Grade at Buckeye and Tray Run, West Virginia, by 1855. Both bridges have been attributed to the work of Albert Fink. The Tray Run Viaduct was 455 feet long, as noted by Charles S. Roberts in *West End: B&O Cumberland to Grafton 1848–1991*, and was replaced with a more-modern metal bridge in 1888, while the massive 443-foot-long stone viaduct at Tray Run today dates from 1907. Plowden notes that the earliest "true metal viaduct" in the United States, which used individual bents and piers, was built in 1868 by Smith, Latrobe & Company for the Cincinnati & Louisville Short Line.

The Erie Railroad appears to have established the basic pattern for subsequent metal viaducts with the construction of its famous Portage Bridge in 1875. Subsequently, Erie was among those railroads to make good use of this type of bridge. Significant to the application of metal viaducts was the advent of cheaply produced rolled iron and steel. This made possible construction of long and often very tall metal viaducts consisting of prefabricated sections assembled on site.

Initially, iron was the material of choice for these bridges. For many years, bridge builders were wary of employing steel, in part because of its greater cost, but also because early steel was softer than iron, which contributed to concerns that it was ill-suited as a bridge-making material. Cheap, commercially manufactured steel became available in Britain when Henry Bessemer patented a process for blowing air through molten cast iron to remove impurities. The Bessemer process was first employed in the United States in 1865. Improved steel-making processes, notably the open-hearth process developed by German-born William Siemens and perfected by French-born Pierre-Emile Martin, soon augmented the Bessemer process. Among the earliest significant applications of steel in American bridges was the Eads Bridge arch in St. Louis, built during 1874 (discussed at length in the next chapter). From that time onward, steel gradually made inroads into railroad-bridge construction, although initially in conjunction with iron components.

The impressive height and length of many tower-supported steel viaducts earned them local fame, even if the engineering behind them was quite common. Somerset Railroad's Gulf Stream "trestle" in central Maine, pictured in this vintage postcard, was constructed in 1904 by the Boston Bridge Company and demolished in 1976. The Gulf Stream Trestle was a fixture in local lore and the topic of stories published by Walter M. Macdougall in *The Old Somerset Railroad*. In its 72 years, the bridge was the site of three deaths. On the card, the viaduct was estimated to be 600 feet long and 115 feet tall, while Macdougall cites a height of 125 feet. *Author collection*

Erie Railroad's second Kinzua Bridge was constructed in 1900 to replace the original 1881 viaduct. The second bridge used riveted steel in place of wrought iron; its lattice supports were also more substantial, measuring 2 feet by 3 feet instead of 1 foot square. Kinzua Bridge was on a line that tapped coalfields in northwestern Pennsylvania. The line was last used for freight in 1957; the line over the bridge was later reopened for excursion service but was closed again for structural reasons prior to its destruction by a tornado in 2003.
Seán Solomon collection

Steel has the major advantage over earlier metal alloys because it is both adequately rigid and resistant to cracks due to its inherent flexibility, and by the 1890s, steel had largely superseded iron for new construction. Prefabricated metal viaducts straddled this transition. The earliest important examples were constructed of wrought iron, while later bridges were entirely steel.

The most common metal viaduct design consists of braced four-legged towers supporting either plate-girder or truss spans, whereby spans are carried both on top of the towers and between them. As

Erie Railroad built this enormous iron viaduct at Portage, New York, in 1875 after its original 1852 wooden frame trestle at this location was consumed in a spectacular fire. This iron viaduct is considered the prototype for the tower-bent support construction that emerged as a popular form in the late nineteenth century for large-span bridges. Where this bridge employs Pratt deck trusses between the towers, many later bridges used plate-girder construction. At 8:30 a.m. on May 16, 1987, westward Delaware & Hudson train symbol EBBU (East Binghamton to Buffalo, New York) eases across the bridge over the Genesee River in Letchworth Gorge at Portage. Norfolk Southern, which now owns the line, has considered replacing this historic structure with a modern bridge, but as of 2007 the 1875 bridge remained in service, albeit much strengthened. *Brian Solomon*

with the later metal truss bridges, prefabricated viaducts were typically designed and installed by commercial bridge companies. Basic components were standard forms arranged to fit specific applications. This type of construction allowed metal truss bridges built to previously unheard of heights and lengths at reasonable prices and in places where the cost of bridging previously had been deemed prohibitively expensive. While cost-effective, these bridges lacked the variety of unique designs of the largest and longest bridges in the early eras of railroad construction. Steel viaducts are characteristic of late-era American construction that used highly engineered, previously unimagined, and comparatively low-grade profiles in the construction of whole new lines that replaced old alignments with reduced gradient and curvature and reached places previous unattainable by railroad.

MONUMENTAL METAL VIADUCTS ON THE ERIE RAILROAD

In 1850, Erie was building its line from Hornellsville (modern day Hornell) to Buffalo, New York, where

In the Conrail era, the former Erie Railroad's low-grade Graham Cutoff via Salisbury Mills, New York, was favored over the original Erie main line via Middletown, New York. The old main line has been largely abandoned since, and all trains, including suburban passenger services, have been diverted to the Graham Cutoff, which is owned and operated by Metro-North. At 7:35 p.m. on May 24, 2000, the photo shows a Metro-North commuter train running from Hoboken Terminal, New Jersey, to Port Jervis, New York, across the Moodna Viaduct at Salisbury Mills, New York. *Patrick Yough*

The Moodna Viaduct at Salisbury Mills, New York, is a 3,200-foot-long steel viaduct that takes its name from Moodna Creek, which runs through the valley. On October 26, 1996, an excursion train led by Chesapeake & Ohio 4-8-4 No. 614 rolls eastward across the tall bridge on its return trip to Hoboken, New Jersey, from Port Jervis, New York. *George W. Kowanski*

the Genesee River Gorge was a significant obstacle. Surveys located the most suitable place to cross this chasm at Portageville, New York. Edward Hungerford in *Men of Erie* (1946) describes the first crossing, the work of the railroad's engineer Silas Seymore, as an enormous wooden bridge that took more than two years to erect. It required an estimated 1.6 million linear feet of timber and 106,280 pounds of iron. It was cleverly designed so that if any one member needed to be replaced, it would in no way compromise the integrity of the structure.

Opened in August 1852 it was among the great wonders of the line: The tracks spanned approximately 900 feet and at its highest point was 250 feet above the Upper Falls of the Genesee (the *American Railway* published in 1889 cites figures of 800 feet long and 234 feet tall). Seymore's work was erased when the bridge was incinerated by fire in spring 1875. Rather than rebuild the massive bridge of wood, Erie's engineers decided upon a metal design that could be erected in much less time and that was unlikely to meet the fate of the first bridge. Using four-legged braced-iron towers—the tallest of which was 203 feet, 3 inches—to support Pratt deck-truss spans, Erie's new bridge spanning 819 feet was in place in just 47 days. As Plowden notes, this bridge established the pattern for many subsequent metal viaducts installed on American railroads.

The bridge has been altered and strengthened on several occasions. Originally, it was single-track, later doubled, only to be singled again. Although the railroad has discussed the possibility of replacing it with a stronger bridge, it remains active today on Norfolk Southern's Southern Tier main line. A severe speed restriction forces all trains to reduce to a crawl when crossing. This has as much to do with the harmonic motion of freight cars at certain speeds as it does the age or frailty of the bridge.

By the early 1880s, Erie had expanded into the coalfields of northwest Pennsylvania. Initially, it had moved coal via a circuitous route involving other railroads. The December 8, 1882, *Railway Gazette* described how Oliver W. Barnes, chief engineer for Erie's New York, Lake Erie & Western affiliate, located a new line to reach the coalfields more directly. To cross the deep gorge at Kinzua Creek, Pennsylvania, Barnes decided on a direct route across the top, which offered the shortest route. In 1881, the railroad awarded Clarke, Reeves & Company of the Phoenixville Bridge Works (of Phoenixville, Pennsylvania) with the construction contract; design engineers were Thomas C. Clark and Adolphus Bonzono.

Work on the bridge began in August of that year and was completed one year later. Twenty towers were constructed, each using four wrought-iron columns of the Phoenix design. The entire bridge spanned 2,053 feet, and rail level was 301 feet above the bottom of the valley, making Kinzua the tallest such bridge in the world at the time of construction. Only 18 years after the bridge was completed, Erie decided to replace it with a similarly designed steel structure that was much stronger, using thicker and more substantial lattice-girder supports in place of the original Phoenix wrought-iron columns. The new bridge, which opened in 1900, was last used by the Erie in 1957. Later, it was incorporated into a state park that operated tourist trains across it. The bridge was no longer in use when it was destroyed by a tornado in 2003.

In the early twentieth century, Erie used tower-bent steel construction for several bridges. The largest was its Moodna Viaduct near Salisbury Mills, New York, constructed around 1909 as part of the 43-mile-long, low-grade Graham Cutoff. According to bridge plans from Metro-North (the New York metro-area commuter agency that now operates the line), Moodna is 3,200 feet long and 200 feet tall.

A construction photo shows the erection of a typical tower-bent steel viaduct at an unknown location. Notice the piers are made from reinforced concrete. The wooden trestle at the left would have been used to deliver construction equipment and materials. Once the piers and lower supports were erected, the higher-level supports and plate girders would have been assembled in place using a rail-mounted crane that worked its way across the structure as it was built. *Author collection*

A Burlington Northern coal train rolls east out of the setting sun on July 14, 1994, and across the Sheyenne River Valley at Valley City, North Dakota. Using standard tower-bent-supported plate-girder construction and opened to traffic in 1908, this bridge was among the most impressive spans on the old Northern Pacific at 3,863 feet long and 162 feet above the river. *Brian Solomon*

Spanning the Des Moines River Valley west of Boone, Iowa, is the most famous structure on the old Chicago & North Western: the awe-inspiring Kate Shelley High Bridge. Built between 1899 and 1901 as part of the Boone-Ogden Cutoff, this bridge replaced the older low-level structure on the original alignment made famous in 1881 by young Kate Shelley, who risked her life to warn a passenger train of a washout. *Railway and Engineering Review* reported that the weight of metal comprising the entire bridge measured 6,080 tons. Today, Union Pacific routinely moves coal trains that weigh twice that amount across the bridge. In mid-2007, UP was in the process of building a new bridge to the north of the existing Kate Shelley Bridge. *Brian Solomon*

A red sunrise silhouettes the steel-girder towers on Chicago & North Western's famous Kate Shelley High Bridge on April 22, 1995, just days before C&NW was officially merged into the Union Pacific. *Brian Solomon*

On July 2, 2000, Montana Rail Link's Gas Local runs over the famous Marent Trestle on the former Northern Pacific line over Evaro Hill, Montana. Located near DeSmet, Montana, this bridge spans 797 feet across Marent Creek, rising to a maximum height of 162 feet above the water. The original structure dates to 1885 but has been rebuilt and reinforced on several occasions to accommodate heavier axle loads. Montana Rail Link was created in 1987 from spun-off former NP trackage in Montana and eastern Idaho. *Mike Danneman*

TALL METAL VIADUCTS SPAN THE WEST

Although first used in the East, tall metal viaducts were prominent features on railroads across the West. In 1882, Santa Fe Railway spanned a deep cleft in the Arizona Plateau known as Canyon Diablo—approximately 26 miles west of Winslow—using an iron tower-bent viaduct supporting plate-girder and Pratt deck-truss spans. As described by Keith L. Bryant in *History of the Atchison, Topeka and Santa Fe Railway* (1974), this bridge was 560 feet long and 222 feet, 6 inches tall. The bridge members were built in New York, shipped to Canyon Diablo, and erected on site. It was later strengthened and finally replaced by a steel arch bridge.

Another early western metal viaduct is the Northern Pacific span of Marent Gulch near DeSmet, Montana, on its crossing of Evaro Hill. Railroad information indicates this uses five 116-foot, 9-inch deck trusses carried on four iron and steel towers with a 30-foot plate girder span atop each tower for a total length of 797 feet and maximum height of 162 feet. Opened in 1885, the bridge has been strengthened on several occasions to accommodate increased axle loads. Today it serves Montana Rail Link.

In 1892, the Phoenix Bridge Company built a 2,180-foot-long, 321-foot-tall iron viaduct across the Pecos River on Southern Pacific's Sunset Route. Plowden cites this as the last significant example of an American iron viaduct. In *Southern Pacific: The Roaring Story of a Fighting Railroad* (1952), Neill C. Wilson and Frank J. Taylor describe the bridge as

A westward Union Pacific double-stack glides over the Clio Viaduct—also known as the Willow Creek Viaduct—on the former Western Pacific west of Portola, California, on the longest day of the year: June 21, 1999. Although the Western Pacific route features several prominent bridges, *Railway and Engineering Review* reported in 1910 that, as built by the American Bridge Company of New York in 1907, there were only 41 steel bridges on the 923 miles between Salt Lake City and Oakland. This is a relatively small number compared with eastern lines. *Tom Kline*

"airy as a cobweb" because of its great height and its comparatively narrow members. At the time, it was the tallest bridge in the United States, 20 feet taller than the Kinzua Viaduct erected by the Phoenix Bridge Company a decade earlier. The Pecos High Bridge was substantially reinforced in 1910 using steel members to allow for greater axle loading. As described in the October 14, 1910, *Railway Age Gazette*, the reinforcing work was designed by J. D. Isaacs, a consulting engineer for the Harriman Lines (Southern Pacific and Union Pacific). The bridge was shortened by 665 feet owing to the construction of a deep fill at the west end. Between 1942 and 1944, Southern Pacific replaced the entire structure with a significantly stronger continuous-steel cantilever bridge with a central span of 374 feet across the river.

Western Pacific was built as the western extension of George Gould's proposed transcontinental system. Between 1906 and 1909, WP extended its line from Salt Lake City to Oakland, California. On the route were several prominent steel tower-bent-supported viaducts, including the Spanish Fork Viaduct at Keddie, California. In its original form this bridge was not especially unusual. According to the March 12, 1910, *Railway and Engineering Review*, it consisted of four steel towers supporting four 30-foot deck spans atop the towers, with 60-foot-long plate girder spans and a 120-foot-long deck-truss span between the towers. At its highest point the rails were 105 feet above the river.

WP constructed its High Line in the early 1930s toward Bieber Station, California, where it met the Great Northern. The westward junction between the new line and the main line was situated at the west end of the Spanish Fork Viaduct, resulting in another leg of the viaduct to carry the new route. The bridge became known as the "Keddie Wye" and has become one of the most photographed bridges in California.

Among California's most photographed railroad bridges is the former Western Pacific Spanish Fork Bridge, better known as the Keddie Wye. It's unusual because the junction of two main lines occurs on a tall plate-girder viaduct; it's popular with photographers because it is easily viewed from California Highway 70. On September 12, 1998, an eastward Union Pacific double-stack container train crosses the main east-west leg of the bridge, while Amtrak's detouring No. 11 *Coast Starlight* holds on the north leg. Amtrak normally operates on the former Southern Pacific Shasta Route, but that line was temporarily closed by a derailment. *Brian Jennison*

The Milwaukee Road, the Spokane, Portland & Seattle, and the Western Pacific were three ambitious projects of the early twentieth century using highly engineered lines with tall, prefabricated steel viaducts characteristic of late-era railroad expansion. A booming economy—combined with improved construction techniques and the ability to build very strong, tall bridges relatively cheaply—allowed railroad construction in places previously deemed cost prohibitive. The shift toward a highway-based economy in the following decades resulted in the eventual abandonment of many of these late-era lines. The Pasco–Spokane section of the SP&S route, seen here at the Burr Canyon Viaduct, was abandoned by Burlington Northern a few months after the photo was made in May 1987. *Brian Jennison*

Milwaukee Road's Pacific Extension was a classic example of late-era construction that employed large steel tower-bent-supported plate-girder viaducts. On August 11, 1979, less than a year before the line was abandoned, freight Extra 21 east crosses the Bandera Trestle in the Washington Cascades. The catenary supports from Milwaukee's famed 3,000-volt direct current electrification were still standing despite deactivation and removal of the wires several years earlier. *Brian Jennison*

CHAPTER 4

Long Spans and Other Large Bridges

SUSPENSION BRIDGES

Of all the primary bridge designs, the suspension bridge is probably least associated with railroads. It would be easy enough to dismiss the type entirely in regards to American railways, or to sum up a short list of suspension bridges that have carried tracks, but to do so would be to omit some of the most interesting and most famous bridges to have ever carried tracks in the United States.

Although various types of suspension bridge had been built for centuries, the design attracted growing interest in the nineteenth century for its ability to span great distances. Between 1820 and 1826, renowned British bridge builder Thomas Telford spanned the Menai Strait in Wales with a chain-supported suspension bridge 580 feet long, at the time the longest single span in the world. This led to a number of significant suspension bridges in Europe, among them the 312-foot chain-suspension span across the Danube Canal in Vienna in 1828 and a famous 870-foot suspension bridge over the Sarine Valley at Fribourg, Switzerland, in 1836.

Right: The monumental masonry towers at each end of Hell Gate feature enormous portal arches for the trains to pass as they cross the bridge. On March 10, 2007, an Amtrak train from Boston crosses the East River. *Brian Solomon*

acela
2002

Detail of the riveted lattice-steel members that comprise the Crooked River Bridge at Terrebonne, Oregon. This splendid steel arch is perhaps the best-known span built for the Oregon Trunk Railway by eminent bridge engineer Ralph Modjeski, who served as the company's chief engineer between 1905 and 1915. A Polish immigrant, Modjeski engineered many prominent North American bridges, including the second Quebec Bridge and the Delaware River suspension bridge. *Tom Kline*

All were built for road traffic, as engineers of the day were still hesitant to use suspension bridges for railway applications.

The importation and adaptation of the suspension bridge to America was neither coincidental nor merely intellectual osmosis. As with the types of exchanges described in Chapter 1, technology transfer was a direct result of American engineers studying bridge design in Europe and bringing their knowledge home. The suspension bridge was adapted in new ways and ultimately grew to greater proportions than anywhere else. Two engineers, in particular, made significant contributions to suspension bridges in America. Charles Ellet Jr., born in Pennsylvania in 1810, was educated in France in the 1830s. In *Engineers of Dreams*, Henry Petroski suggests that in Ellet's European travels he may have been inspired by the Sarine Valley and suspension bridges of the pioneering French builder Marc

Oregon Trunk's famous steel arch spanning the Crooked River Gorge at Terrebonne, Oregon, is 350 feet long and 320 feet high (some sources say 340 feet). Opened in 1911, its late construction was a product of rivalry between the Hill and Harriman railroad empires in the Pacific Northwest. On September 12, 2003, a southward BNSF freight makes its way across the bridge. *Tom Kline*

The longest American railroad span is the Huey P. Long Bridge over the Mississippi at New Orleans. Completed in 1935 at an estimated $13 million, the total structure is 22,997 feet long, with the span across the river measuring 790 feet. Like many remarkable U.S. bridges, it was used by Southern Pacific. This bridge, like the smaller Huey P. Long Bridge at Baton Rouge, carries both a road and railroad. *Author collection*

A westward Santa Fe train rolls across Arizona's Canyon Diablo steel arch on April 19, 1990. This spandrel-braced cantilever arch spans a chasm 222 feet deep, and was completed in 1947, making it the third bridge at this location. The piers for the earlier steel viaduct can be seen in the canyon. A spandrel-braced arch forms a very rigid structure, which makes the design especially desirable for large main-line railroad bridges. *Mike Danneman*

On January 20, 1996, a Kansas City Southern freight crawls up the west approach to the Huey P. Long Bridge at Baton Rouge, Louisiana. Like its larger and more famous cousin at New Orleans, the Baton Route bridge serves both a highway and railroad; in this case, Airline Highway U.S. Route 190 and KCS with Union Pacific trackage rights. Since railroad operations require a gentler gradient than highway traffic, railroad approaches are much longer. The approach structure uses standard steel-tower-supported plate-girder viaduct construction. *Brian Solomon*

Seguin, which are believed to be predecessors to Ellet's U.S. projects. Sigfried Giedion, in *Space, Time and Architecture, 4th Edition.* (1963), describes the wire-cable bridge built by Seguin across the River Rhone at Tournon in 1824, and David Steinman and Sara Ruth Watson, in *Bridges and Their Builders* (1957), allude to an early railway application of a Seguin wire suspension bridge over the River Saône. Completed by 1840, the latter measured 137 feet long and incorporated stiffening girders.

In 1842, Ellet designed and erected a road bridge across the Schuylkill River at Fairmount Park in Philadelphia that Petroski and others cite as America's first wire-cable suspension bridge. Then, in the mid-1840s, Ellet simultaneously worked on two significant projects. At Wheeling, Ohio, he built a 1,010-foot suspension bridge across the Ohio River—at the time the longest span of any bridge in the world. At Niagara Gorge, he competed for the contract to a similar venture that was to carry a heavy railway line and a road. Another competitor for the contract was John A. Roebling.

Roebling is arguably the most famous of all American bridge builders. In his day, he was considered a genius of nineteenth-century bridge engineering. Along with his son Washington, Roebling designed the world-famous Brooklyn Bridge across the East River in New York. As a result, the family name is synonymous with suspension bridge design.

Born in Prussia in 1806, Roebling studied architecture and engineering in Berlin and then immigrated to America, where, by the early 1840s, he was offering opinions on bridge designs and as early as 1841 had become a proponent of the suspension

A Southern Pacific freight led by Norfolk Southern C39-8 No. 8623 crosses the steel-tower-supported plate-girder deck approach at the east end of New Orleans' Huey P. Long Bridge at Jefferson, Louisiana, on October 20, 1990. Prior to construction of the Long bridge, SP reached New Orleans with car ferries. This bridge uses very long and expensive approach viaducts to allow the bridge to meet minimum span requirements for height and length over the Mississippi. The main span uses a cantilever truss with 145-foot arms and a 500-foot suspended section providing a minimum of 135 feet of clearance over the river. *Mike Abalos*

bridge. By 1844 he was designing and building suspension bridges for use by canals and aqueducts. Converted to road use, his Delaware River suspension bridge at Lackawaxen, Pennsylvania, survives to the present day.

NIAGARA FALLS BRIDGE

Best known for its tremendous falls, the rapidly flowing Niagara River connects Lake Erie with Lake Ontario, and it separates New York State from the Canadian province of Ontario. The difficult combination of the very deep and wide Niagara Gorge and the ferocious nature of the river precluded conventional bridging methods at this key location, where there was great need for a railroad bridge. A suspension span appeared to be the ideal choice. However, the project was notoriously controversial, and many prominent engineers, including Robert Stephenson, argued against using suspension bridges for railroad applications.

Initially, Ellet won the favor of the Niagara Bridge Company. Steinman and Watson tell of his success in constructing a 770-foot suspension footbridge across the gorge in 1847 and 1848 as the prelude to a more-substantial structure. Ironically, this bridge proved his undoing. Soon after this span's completion, Ellet and the Niagara Bridge Company came to an impasse on the application of tolls, which resulted in Ellet's resignation. Roebling was hired in Ellet's place and between 1851 and 1855 constructed a magnificent bridge that, as described by Middleton in *Landmarks of the Iron Road*,

Above: The Niagara Gorge suspension bridges were popular with producers of commercial stereograph images. The three-dimensional stereopticon system was invented by the poet Oliver Wendell Holmes and made its world debut at the Crystal Palace in London in 1851. The medium was an instant success in Britain as a result of Queen Victoria's endorsement. Roebling's cable-suspension bridge not only made for an intriguing three-dimensional subject—like the stereograph, it was representative of Victorian "high technology." *Author collection*

spanned 821 feet across the frothing rapids. This double-deck structure carried one track on top and a roadway on the bottom. It was supported by wire-spun cables suspended from massive, 60-foot-tall limestone towers on opposing rims of the gorge and firmly anchored to bedrock beyond.

Key to the success of Roebling's design was the trussing between the decks designed to stiffen the bridge against the load of trains passing over it, as well as the fierce winds that blew up the gorge. Roebling's bridge was a sight to behold and soon became a tourist attraction in its own right. There was nothing else like it in North America nor in the world. While the bridge helped propel Roebling's reputation and his career, and helped establish the wire-cable, long-span suspension bridge in North America, it was to remain an anomaly in heavy North American railroading.

In 1897 and 1898, Roebling's famous Niagara Suspension Bridge was replaced by this huge steel arch. It was designed and built by Leffert L. Buck, who went on to design the Williamsburg Bridge in New York City. This view was made from almost the same angle as the Victorian stereograph. In both, a train is crossing the bridge from Canada. In this image, Amtrak F40PH 326 leads the *Maple Leaf* from Toronto to New York City in November 1988. The Michigan Central (New York Central) arch to the left of the train was built in 1924–1925 to replace a steel cantilever bridge on that line. *Brian Solomon*

Racing along on the outer edge of the Benjamin Franklin Bridge is a third-rail rapid transit train en route from 8th and Market, Philadelphia, to Camden, New Jersey, in 1961, before the line became the Haddonfield PATCO route. The bridge, which opened as the world's longest single span in 1926, held this title for just three years. It has carried rapid transit trains since June 1936. *Richard Jay Solomon*

John A. Roebling's posthumous masterpiece was the Brooklyn Bridge, completed with the aid of his son, Washington Roebling, and daughter-in-law, Emily Warren Roebling, and opened in 1883. Although it did not carry heavy main-line trains as did the Niagara Suspension Bridge, the Brooklyn Bridge hosted both rapid transit trains and electric streetcars until it was reengineered and rebuilt by David Steinman in the 1940s. The El tracks were removed in 1944. *Richard Jay Solomon collection*

Leffert L. Buck was chief engineer of New York's Williamsburg Bridge. Previously Buck had engineered repairs on the Niagara Suspension Bridge and later built the steel arch that replaced it. The Williamsburg Bridge opened in 1903 and carried electric trolley cars. Later, it carried elevated rapid transit trains and, for a short time, Long Island Rail Road electric multiple-units. *Richard Jay Solomon collection*

However, many of the great suspension bridges built later did have tracks for rapid transit and trolley lines, including Roebling's Covington (Kentucky) and Cincinnati (Ohio) suspension bridge (1866), which anticipated his Brooklyn Bridge design; the three great East River crossings in New York City—Brooklyn Bridge (1883), Williamsburg Bridge (1903), which, as noted, once carried Long Island Rail Road suburban trains, and Manhattan Bridge (1909)—the Benjamin Franklin Bridge (1926) between Philadelphia and Camden, New Jersey; and the TransBay Bridge (1936) between San Francisco and Oakland. All are still in use, although the Cincinnati, Brooklyn, and TransBay bridges no longer have rail transit (though there have been proposals to put rails back on the TransBay).

The Niagara Falls suspension bridge served for more than 40 years. It was eventually acquired by the Grand Trunk (later a component of Canadian

Leffert L. Buck blended cable suspension and cantilever forms in his designs. In its day, the Williamsburg Bridge was decried as unsightly because its raw girders and Spartan functional structure were seen as the antithesis of the graceful lines of Roebling bridges. Today, it is considered one of the classic East River spans. It is seen from the Manhattan side of the river on March 10, 2007. *Brian Solomon*

National Railways), but suffered the fate of a great many railroad bridges built of the mid-nineteenth century: it was ill-suited to accommodate much heavier trains that came later. Petroski notes that Leffert L. Buck, who became the chief engineer for New York's Williamsburg suspension bridge, oversaw repairs of the Niagara span in the 1880s and then designed and initiated construction of the immense steel arch that replaced it in 1897.

EADS BRIDGE

Among the most significant long bridges of the later nineteenth century was the Mississippi crossing at St. Louis designed by James Buchanan Eads. Although not the earliest crossing of America's largest river, Eads' bridge set several important precedents as a distinct product of nineteenth century engineering. Its construction is fascinating due somewhat to the character of Eads himself, who, unlike modern bridge engineers, was largely self-taught.

Bridging the Mississippi was unusually difficult. The river is exceptionally wide and has exceptionally swift currents that carry large quantities of silt and debris. In order to sink foundations for the bridge, it was necessary to penetrate up to 80 feet of silt and sand that was up to 25 feet below river level. Overcoming the technical hurdles of building the bridge was only part of the problem—the bridge itself was fiercely controversial. Mississippi riverboat interests viewed railroad transport as a serious threat and claimed piers in the river would present dangerous obstacles. They were vehemently opposed to the project and politically well-connected. In order to keep the river open to steamboat navigation, the bridge required broad

The Benjamin Franklin Bridge across the Delaware River at Philadelphia is among the American long-span suspension bridges to carry rapid transit. This silhouette of the bridge against a stormy sky on October 28, 2006, reveals a four-car Port Authority Transit Corporation train en route from Haddonfield, New Jersey. The main span over the Delaware is 1,750 feet, and the total length of the bridge is 3,183 feet, 4 inches. The bridge was designed to carry streetcars, as well, and a terminal was built for them on the Philadelphia end, but the service was never initiated. *Brian Solomon*

spans with sufficient vertical and lateral clearance for riverboats of the period.

Construction began in August 1867. Work on the piers was started using cofferdams—the established method for sinking bridge piers. In 1869, with construction underway, Eads traveled to Europe, where he learned about a new technique for sinking piers using pneumatic caissons, pressurized vessels allowing work underwater. Eads was well familiar with underwater work, having spent much time in diving bells recovering shipwrecks. He returned with the concept of the pneumatic caisson, which made possible the sinking of bridge piers to the required depths. Although he didn't invent caisson-working, Eads Bridge has been considered the most significant pioneer application of caisson technique in the United States, setting precedents for later construction. An unfortunate part of the learning experience included numerous fatal encounters with what was then described as "caisson disease," later known as decompression sickness or the bends. As a result, Eads encouraged pioneering research into the ailment that led to improved understanding of the problem and better safety precautions.

Another significant precedent was Eads' choice of construction material: steel. Commercially produced steel was new to large construction, and its properties were not yet fully appreciated. Eads decided on cast steel for the arch spans because it could carry a heavy load without yielding in compression and was well suited to anticipated levels of

Completed in 1874, Eads Bridge at St. Louis carried main-line trains on its lower level for 100 years. Today, it hosts St. Louis' MetroLink light rail. *Scott Muskopf*

expansion and contraction resulting from temperature extremes in St. Louis. The bridge was built using three huge elliptical arches: two outside spans measuring 502 feet and a central span measuring 520 feet. Each span consisted of four steel arches fabricated from sections of tubular structural steel 18 inches in diameter and 12 feet long. These parallel arches were spaced 12 feet apart and fastened together with diagonal bracing.

The last of the steel arches were completed by December 18, 1873, and the bridge opened for service in July 1874. It has two levels: the top carries a roadway, and the bottom is a double-track railroad originally used by the several companies serving the St. Louis Gateway. Eads Bridge continued to carry commercial railroad traffic for 100 years, but the clearances on its lower level designed for standard railroad cars of the 1870s were too small to accommodate many modern freight cars, and it was last used by conventional railroad traffic in 1974. The bridge was later adapted for St. Louis' new light-rail transit system, MetroLink, which opened in 1993.

POUGHKEEPSIE BRIDGE

In the wake of the Civil War, business interests in the Poughkeepsie, New York, area envisioned a trans-Hudson span at their town and helped form the first bridge company to consider the project in 1871. According to Union Bridge Company's J. F. O'Rourke, who prepared a detailed paper on construction for the American Society of Civil Engineers that was reprinted in the July 1, 1887, issue of *Railroad Gazette*, the original cornerstone was set in December 1873, implying that work on the bridge began prior to January 1874. In actuality, as Middleton points out, the first bridge company lacked financial resources, and the project stalled for more than a decade until 1886, when the Manhattan Bridge Company stepped in and contracted the New York City–based Union Bridge Company to finish the work. Among the greatest difficulties in building the Poughkeepsie Bridge was establishing piers in the Hudson River; this is where initial construction had floundered.

Union Bridge's chief engineer was Thomas C. Clark, a Harvard University graduate of 1848 who had worked on a variety of bridge projects, including the Kinzua Bridge (see Chapter 3). The *Railroad Gazette* article noted that, in addition to piers on either side of the river, four mid-river masonry piers were necessary. These were sunk deep into the riverbed (135 feet below the high-water mark) and rose 30 feet above the water.

The Eads Bridge over the Mississippi River at St. Louis was among the most significant spans of the nineteenth century. Its construction saw pioneering use of pneumatic caissons for its mid-river piers. *Voyageur Press collection*

The Eads Bridge's arches were among the first significant applications of steel on a large bridge. *Author collection*

The central spans of the bridge are of classic cantilever design. O'Rourke noted these specifications and statistics about the bridge as it neared completion:

> *The total length of the Poughkeepsie Bridge will be 6,667.25 ft. The west approach is 1,033.5 ft., being a viaduct of two pin connected trusses of 145 ft. each, 7 latticed girders of 60 ft. each, one of 53 ft., and nine plate girders of 30 ft. each. The total length of the bridge proper is 3,093 ft. 9 in., consisting of two shore spans (the shore arms of the cantilevers) of 200 ft, 101/2 in. each, and five river spans from 525 to 548 ft. each. These spans are cantilevers with connecting spans. The east approach is 2,640 ft., being a viaduct of one truss of 175 ft. span, one of 161 ft., three of about 116 ft. each, and the rest latticed and plate girders of from 30 to 85 ft. each.*
>
> *The elevation of the base of rail above high water is 212 ft. and the head-room is from 130 ft.*

The Poughkeepsie Bridge was more than just a crossing of the Hudson–it was a key gateway for freight moving to and from New England. Built using pneumatic caissons, each mid-river pier is 89 feet, 6 inches long, and the above-water segments measure 24 feet, 6 inches wide. The bridge is 212 feet from rail top to mean water level. *Author collection*

When Illinois Central's Ohio River Bridge opened on October 29, 1889, it formed the final link in the railroad's north-south main line from Chicago to New Orleans. It was the longest bridge in the United States; Carlton J. Corliss, author of *Main Line of Mid-America*, wrote that it was 20,461 feet long, including approaches. Rail level was 104 feet above low water. The spans depicted consisted of long Whipple trusses mounted on masonry piers. George S. Morrison was the chief engineer, and Ralph Modjeski served as an inspector of superstructure. *Author collection*

The Poughkeepsie Bridge looms like a skeletal leviathan above a CSX freight operating on the former New York Central West Shore Route. On March 8, 2001, the bridge was just a sad reminder of better times on the old New Haven. Plans are underway to convert the enormous bridge into a footpath. In all likelihood, it will not carry trains again. *Brian Solomon*

to 163 ft. The river piers from 30 ft. above high water are latticed towers. The structure is entirely of steel, even to the rivets, excepting some portion of the viaduct approaches.

The Poughkeepsie Bridge was completed at the end of 1888. The first train to cross it was hauled by a Hartford & Connecticut Western locomotive on December 29 under the able hands of engineer George Austin, as recalled by Robert Ashman in *Central New England Railroad* (1972). The Poughkeepsie Bridge was operated as part of the Central New England & Western system, later the Philadelphia, Reading & New England, reorganized as the Central New England Railway after 1898. CNE came under the wing of the New York, New Haven & Hartford Railroad after 1900 and was finally merged into the New Haven in 1927. The New Haven developed a new southerly route to the bridge via Danbury and Devon, Connecticut, that allowed classification of traffic at its vast Cedar Hill Yards east of New Haven. After New Haven was absorbed by Penn Central in 1969, declining traffic was largely detoured via former Boston & Albany routes by way of Selkirk and the Castleton Cutoff in New York State. When the Poughkeepsie Bridge deck caught fire on May 8, 1974, it had been accommodating only a single daily roundtrip. Penn Central, by then bankrupt, had little incentive to repair the damage. The bridge has not carried a train since. Walkway Over the Hudson, a nonprofit group that hopes to reopen the bridge for pedestrians and cyclists, acquired the bridge in 1998.

Each of the three massive through trusses on Santa Fe's Missouri River Bridge spans 396 feet. Completed in 1915 to replace Santa Fe's original 1887 structure, the bridge was built with a creosoted and ballasted deck, which, according to *Railway Age Gazette*, was unusual for long truss spans at the time. This February 2007 view depicts the structure as it now serves BNSF Railway. It is one of three single-track segments remaining on the railroad's heavily trafficked Chicago–California "Transcon." *Dan Munson*

SANTA FE'S MISSOURI RIVER CROSSING

In 1887, Santa Fe Railway decided to extend its line to Chicago. The terrain between Kansas City and Chicago is mostly rolling prairie, and Santa Fe was able to survey an "airline" route, characterized by long tangents, gentle curves, and no serious grades. However the Missouri River crossing at Sibley, Missouri, roughly 25 miles east of Kansas City, proved a critical obstacle. As reported in the July 2, 1915, issue of *Railway Age Gazette*, at this location, the railroad built a 4,082-foot-long, single-track bridge. On the eastern end, a 2,000-foot, tower-supported plate-girder viaduct was used, followed by three deck-truss spans, two measuring 172 feet 6 inches, and one at 247 feet long. The

This hand-tinted postcard depicts Santa Fe's new bridge over the Missouri River that was completed in 1915. The back of the postcard boasted: "Experts and engineers have come from all parts of the country and from Europe to see the Santa Fe bridge at Sibley, Mo. The bridge was built on enlarged piers of the old structure and the old bridge removed without delaying traffic a single minute." *Author collection*

In October 1995, an eastward BNSF freight led by Santa Fe GP60M 127 rolls across the Missouri River Bridge at Sibley, Missouri. Note the especially deep deck truss and pin-connected members. The three through trusses are each 396 feet long, the two plate-girder spans resting on masonry piers nearest the through trusses on the east end of the bridge are each 99 feet, 3.5 inches long, and the span nearest the east bank of the river is 80 feet, 4 inches long. Plate-girder spans between tower bents on the westward approach are 90 feet long, and those over the tops of the towers are 45 feet long. *Chris Guss*

navigable channel was spanned by a trio of through trusses, each 396 feet long, and the western section of the structure required a 200-foot deck truss, plus an 80-foot plate-girder deck.

When service to Chicago began in 1888, Santa Fe had the only through route from Chicago to California under the control of one company. In the years prior to World War I, Santa Fe designed and built a far more substantial replacement bridge on essentially the same alignment, although slightly longer at 4,100 feet between abutments. Work on the new bridge was directed by Santa Fe's chief engineer, C. F. W. Felt, and a bridge engineer identified as Robinson (it is unclear whether this is the Albert A. Robinson identified in other historical sources), with help from divisional engineer G. J. Bell and assistant engineer F. H. Frailey. Steel members were manufactured in Gary, Indiana, by the American Bridge Company, and much of the erection was completed by the Missouri Valley Bridge and Iron Company. Work began in 1911 and was completed by 1915.

The new structure used a ballasted deck throughout and allowed for a gauntlet arrangement in place of true single-track. On the east end the new viaduct consisted of 45-foot-high towers supporting 90-foot intermediate spans, while the main spans in the middle were of the same length as those used on the first bridge. On the west end, the 200-foot deck truss was replaced by a pair of 100-foot plate-girder deck spans. The new bridge reduced the grades on approaches from 0.8 percent

Restored Northern Pacific Ten-Wheeler No. 328 leads an excursion train across the former Soo Line bridge over the St. Croix River near Somerset, Wisconsin, on August 17, 1996. At this time, the railroad was operated by Wisconsin Central Limited, a spin-off regional railroad established in 1987, using much of the traditional Wisconsin Central network that had been a component of Soo Line. This bridge was built by Soo Line in 1910–1911 as part of a 17.5-mile cutoff to improve the connection between Soo Line's main line (from Minneapolis and Sault Ste. Marie, Michigan) and WC's route to Chicago. Each of the five cantilever arches spans 350 feet; the bridge, as built, was 2,682 feet long, and in 1911 the base of rail to the high-water mark was 169 feet. *Brian Solomon*

to 0.5 percent by raising the level of fills leading up to the bridge, which reduced operational difficulties for heavy trains. The bridge remains in service today and is one of just three single-track sections left on the largely double-track BNSF Railway transcontinental main line.

ST. CROIX RIVER STEEL ARCH

In 1910–1911, the Minneapolis, St. Paul & Sault Ste. Marie—the Soo Line—planned and built a 17.5-mile, low-grade cutoff connecting its Portal, North Dakota–Sault Ste. Marie, Michigan, main line with its Wisconsin Central main line to Chicago. New trackage ran from Withrow, Minnesota, to New Richmond, Wisconsin, crossing the St. Croix River Valley near Arcola Station (north of Stillwater, Minnesota) on a massive steel arch.

As described in the December 1, 1911, *Railway Gazette,* the line had a maximum gradient of just 0.5 percent in comparison with the more circuitous line it replaced, which had a maximum grade of 1.3 percent. The most significant feature of the cutoff was the St. Croix Bridge, designed by consulting engineer Claude A. P. Turner and constructed between autumn 1910 and June 1911 by Thomas Green and his assistant, C. N. Kalk. Steel viaducts using tower-supported plate-girder construction were used on both approaches to the main span across the St. Croix, the east-side viaduct measuring 340 feet long, the west side 560 feet.

These viaducts were typical for the period. Unusual were the main spans that consisted of five great arches, each spanning 350 feet and rising 159 feet above the tops of the concrete piers. Rail level was estimated at 169 feet above high water, and the whole bridge spanned 2,682 feet. A tower-supported plate-girder trestle could have been built for the full length of the bridge had the engineers decided entirely on this type of construction. Such a bridge would have required less steel and would have been easier to design and build, but the novel steel arch was adopted to reduce the number of piers, which, because of the unusual depth required to reach bedrock, were exceptionally costly. The bridge remains in service after almost 100 years of continuous service and today serves Canadian National.

HELL GATE

Gustav Lindenthal was born in Brünn in the Austro-Hungarian Empire's Moravia—today's Brno in the Czech Republic. He immigrated to the United States in 1874 and became one of the most famous American bridge builders, perhaps rivaled only by Roebling in the public's consciousness. In 1902 he became bridge commissioner for New York City, putting him in the powerful position of overseeing the great crossings of the East and Harlem rivers.

During the late nineteenth century he established his reputation as master bridge builder, first by erecting a variety of distinctive structures in Pennsylvania. Among these were replacements of a number of worn-out Howe trusses dating from the mid-1800s. George H. Burgess and Miles C. Kennedy note in the *Centennial History of the PRR* (1949) that in 1884 Lindenthal initiated discussions with that railroad regarding his trans-Hudson suspension bridge, discussed above.

In 1900, Pennsylvania Railroad acquired controlling interest in the Long Island Rail Road as part of their strategic plans to improve their New York–area connections, including direct connections to New England via the New Haven Railroad. After weighing a variety of options for crossing the Hudson and East rivers, the railroad finally decided on tunnels under both rivers for its lines feeding its new Manhattan terminal. Lindenthal, who had worked as a consulting engineer for PRR since 1904, was hired for design and construction of a connection with the New Haven Railroad via the PRR subsidiary New York Connecting Railroad by building a bridge across "Hell Gate," the 850-foot-wide tortuous river estuary whirlpool between the boroughs of Queens and the Bronx.

Lindenthal's massive Hell Gate arch has become a familiar sight to New Yorkers, and it's difficult to

Gustav Lindenthal's masterpiece in steel: New York's Hell Gate Bridge. Ninety years after its completion, this magnificent bridge remains an integral part of the region's railway network. It is seen against the evening sun on March 9, 2007. In the distance are New York's Tri-Borough Bridge and the Manhattan skyline. *Patrick Yough*

Hell Gate Bridge is among the most impressive railroad bridges in North America. It was the inspiration for John Bradfield's Sydney Harbor Bridge in Australia, which, like Hell Gate, carries tracks but also a roadway. Sydney Harbor Bridge carries a nominally heavier deadweight, but Hell Gate is a greater load-bearing structure. On June 22, 1958, a streamlined New Haven EF-3 electric leads a work train across Hell Gate's arch span. Notice the long eastward approach viaduct to the Bronx visible below the arch. *Richard Jay Solomon*

imagine any other type of bridge at this location, yet an arch was not initially considered for the crossing. Among the requirements for a bridge, as set by the United States War Department, was that it leave an open channel at least 700 feet wide and clear a minimum 135 feet in height. Lindenthal, long a proponent of suspension bridges, considered this design, along with a three-span continuous cantilever and a continuous three-span truss. Arguments against an arch cited the lack of natural geological abutments for support. Despite opposition, Lindenthal ultimately chose an arch to meet both economic criteria (Petroski indicated in *Engineers of Dreams* that the arch offered lower material costs) and his personal aesthetical concerns. He envisioned Hell Gate as the water gateway to New York via Long Island Sound, and so it demanded a pleasing and impressive design. Othmar H. Ammann and David B. Steinman were Lindenthal's chief assistants for design and construction; both went on to design some of America's most famous twentieth-century spans (Steinman also co-authored, with Sara R. Watson, *Bridges and Their Builders*). Construction on the bridge was initiated in 1911, and much of the heavy work was completed between 1914 and 1916. The bridge opened to traffic in early 1917 with a formal dedication on March 9 of that year.

Keys to the arch design are the monumental masonry towers—250 feet tall—on both ends of the span. These serve as the abutments for the bottom chord, transmitting its forces to the ground. Although the height of these towers was largely a result of Lindenthal's aesthetic considerations, their great weight is required to buttress the principal load of the span. The span consists of a double-hinged parabolic arch with supporting variable-depth trusses connecting the lower chords to arched top chords. The truss ranges from 190 feet at the ends to just 40 feet at the apex of the arch. The top of the arch span is 280 feet above mean tide, and the four-track ballasted track bed is suspended from the arch with vertical steel members consisting of 23 panels; it is the only long railroad span in the world to have had four tracks, separated in pairs for passenger and freight, though one freight track has since been removed. A New York Connecting Railroad drawing of the bridge dated March 28, 1917, indicates that the bottom of the deck clears the river by 135 feet (measured above mean high water). Although the bridge was essentially a Pennsylvania Railroad project, operationally it served New Haven's freight and passenger trains. Today the route is key to Amtrak's Northeast Corridor trains between New York and Boston.

Another Lindenthal masterpiece was completed the same year as Hell Gate: Chesapeake &

The rising sun over Suisun Bay in August 1993 silhouettes Southern Pacific's 1930-built double-track, multiple-span truss, as well as the more modern I-680 bridge that runs parallel to SP's bridge. In the foreground are the rails of Southern Pacific's line to Oakland, California, along the southern shore of Suisun Bay. This photo shows four of the Warren through truss spans in profile. Each measures between 529 feet, 6 inches and 531 feet long. At the right is SP's vertical-lift span, 336 feet long. Since this photo was made, a second massive highway bridge was built to the east of SP's bridge. *Brian Solomon*

Ohio's Ohio River span at Sciotoville, Ohio. This was significant for its use of a 1,550-foot-long, continuous Warren-style truss using two spans of 775 feet each, a design similar to what he had contemplated for Hell Gate.

SUISUN BAY BRIDGE

Central Pacific's transcontinental line via Donner Pass was completed to Sacramento, California, in 1869. Later that year, the railroad reached Oakland, California, opposite San Francisco, by way of a circuitous routing via Tracy and over Altamont Pass. Another circuitous route to Oakland following a water level profile via Tracy and Martinez was opened in 1878. A direct line running from Sacramento to Oakland finally opened 1879, but required a car ferry across Suisun Bay between Benicia and Port Costa, a few miles west of Martinez on the 1878 route. The car ferry *Solano* was considered the largest of its kind in the world. Although a bridge was considered for this crossing from the beginning, at the time the cost of such a span was deemed prohibitive. When the line was doubled in 1914, Southern Pacific Company (which included Central Pacific), added a second large car ferry, *Contra Costa*.

Fifty years after the line opened, SP finally decided to bridge Suisun Bay and began work on the massive, multiple-span crossing. SP's engineer for maintenance of way and structures, William H. Kirkbride, presented a paper to the American Society of Civil Engineers in April 1924, abstracted in the June 21, 1930, *Railway Age*, that detailed the history and construction of the bridge. Among the special considerations was the potential for earthquakes, since it was close to a significant fault line. Designed to resist tremors up to twice as severe as the quake that destroyed San Francisco in 1906, the bridge features concrete piers constructed with material capable of withstanding high compressive and tensile forces, and located with as low a center of gravity as practical. Also of special consideration was construction of piers in the deep and rapidly flowing water of the Carquinez Straits. New methods for erecting

In its day, Southern Pacific's enormous Suisun Bay Bridge was an awe-inspiring structure. On a 1955 afternoon, one of SP's oil-fired cab-ahead steam locomotives leads an eastward freight over the bridge. The gradient up the eastward approach viaduct was a conservative 0.6 percent. While mild compared with the brutal 2.4 eastward ruling grade over Donner Pass, this nominal climb slowed heavy trains to a crawl. Since the rest of the route between Oakland and Roseville, California, was virtually level, as late as the early 1950s, freights sometimes required helpers. The big articulated steam locomotive is dwarfed by the 266-foot-6-inch-long deck span that it is crossing–the first of 10 trusses on the bridge. *Fred Matthews*

piers were developed that required construction of a temporary artificial island.

The bridge was built for double track to very high loading standards for the day. On both sides of the bay the approaches used typical steel-tower-supported plate-girder viaduct construction, bringing the tracks up to the level of the main span. The entire bridge, including approach viaducts, is 5,603 feet, 6 inches long. Principle members were made from silicon steel, with carbon steel used for most other components. A total of 22,000 tons of steel was used in the superstructure. The bridge was completed at an estimated cost of $12 million and opened for service on October 15, 1930. It remains a principal crossing on Union Pacific today, handling dozens of freight and Amtrak passenger trains daily. It is on the route of Amtrak's *California Zephyr*, *Coast Starlight*, and *Capitols*.

6397
CONRAIL QUALITY
6397
CONRAIL
CR
50 74 21

CHAPTER 5

Replacement Spans

The story of North American railroad bridges is one of ever-longer, stronger, and more-durable structures. This was necessary to meet the demands of heavier and faster trains, provide greater levels of safety, and require less maintenance. The railroads' early wooden trusses built in the industry's formative years became obsolete or wore out within a few decades. They were replaced by a variety of quirky iron trusses, including those of Bollman, Fink, and Whipple patents, as well as a host of less-durable designs. By the end of the nineteenth century these iron bridges could no longer withstand the heavier axle loading and greater traffic of modern railroading, and they required replacement with even more durable spans using improved steel trusses, steel-girder trestles, plate girders, masonry or concrete arches, and often a combination thereof as bridge designers saw fit.

Likewise, lines built in haste with cheap wooden trestles had to be improved. Often, the old trestles, if not filled in with stone, earth, and ballast,

Left: Pennsylvania Railroad built the immense double masonry arch at Mineral Point, Pennsylvania, as part of a general upgrading of its Main Line in the late nineteenth century. On September 5, 1997, a quartet of Conrail SD40-2s shove on the back of a coal train. This bridge is unusual not just because it was built to handle four tracks (many such stone arches were built to this specification on the PRR), but because it is located on a sharp curve and steep grade. *Brian Solomon*

On June 10, 1973, a Lehigh Valley freight gingerly negotiates a pier-damaged, double-track truss over the Chemung River at Athens, railroad-direction east of Sayre, Pennsylvania. *R. R. Richardson photograph, Doug Eisele collection*

were replaced with more substantial iron or steel bridges.

In the first decades of the twentieth century, the railroads' golden age, locomotives and trains reached proportions unimagined by the early railroad builders. Yet, labor and materials were still comparatively inexpensive—inflation had not become endemic, and severe governmental regulation had not begun to discourage investment. In this positive economic environment, railroads embarked on massive rebuilding plans, during which they installed many of the bridges that have survived to the present day. New bridges were only part of the story; new lines were built bypassing not only the old bridges but traditional alignments as well. These new cutoffs had fewer and gentler curves, high embankments, and deep cuts, and they often required massive bridges to span whole valleys. When observing these later rights-of-way it is often possible to discern the location of the older lines they superseded. In the shadow of monumental modern spans often lurk traces of the line's early history: a bit of an old fill may be evident lower in the valley, or a row of masonry bridge piers—sans bridge—can be found in a riverbed.

Interestingly, these modern low-grade lines were products of boom times that occurred on the eve of the unforeseen highway age, when it appeared that railroad traffic would continue to rise indefinitely, and new and more efficient operations

The former Great Northern truss over the Columbia River at Rock Island, Washington, has been reinforced over the years to safely accommodate greater axle weights. On October 1, 1994, an eastward train rolls across the distinctive structure. *Tom Kline*

would grab profitable business from older, established railroads. Ironically, these heavily built, late-era lines were often the first to suffer during the age of consolidation, from the mid-1950s to mid-1980s, when railroads sought to reduce their overall route structures. Lines such as the Western Maryland's Connellsville Extension and Spokane, Portland & Seattle's Highline between Spokane and Pasco were abandoned, modern bridges and all, in favor of more-traditional routes.

While bridge replacement has continued to the present, modern bridges rarely resemble those from the age of steam; today's structures are often made of reinforced concrete or steel girders and as a result resemble bridges built for contemporary highways.

PENNSYLVANIA RAILROAD BRIDGE REPLACEMENT

In 1887, PRR began replacing wooden bridges with solid masonry arches on its east-west Main Line. Then about 1900, PRR began reconstruction of its heavily traveled Philadelphia–New York City route, investing a large sum for replacement bridges, largely of masonry construction. In 1902, *The Railway Magazine,* published in London, described PRR's project, then a work in progress, as expected to cost an estimated $20 million. Among the most

As reported in the March 7, 1925, *Railway Review*, in 1883, Michigan Central built a twin-tower steel cantilever bridge adjacent to the famous Niagara Suspension Bridge. This structure was 910 feet, 2 inches long and featured a 119-foot, 9-inch suspended deck truss over the center of the gorge. Designed by C. C. Schneider, the cantilever was effectively obsolete by the turn of the twentieth century, due to dramatic increases in axle weights and trains speeds. The steel arch pictured here was built during 1924 and 1925 as a replacement span. The total length of the bridge, including reinforced concrete approach spans, is nearly 915 feet. From the base of the rail to the low water mark is 240 feet. In November 1988, CSX's daily freight running west from Buffalo to Detroit eases across the bridge. *Brian Solomon*

A Norfolk Southern freight rolls west across the concrete arch bridge over the Conemaugh River at Johnstown, Pennsylvania, on November 3, 2001. The 1887-built elliptical arch bridge that once stood here was made famous when on May 31, 1889, the South Fork Dam burst, sending a tidal wave of destruction down the valley to Johnstown. The bridge withstood the deluge of water and debris, which is said to have accumulated 15 feet above track level. *Brian Solomon*

significant replacement spans were massive four-track bridges over the Raritan River at New Brunswick, New Jersey, consisting of 18 arches, and across the Delaware River at Trenton, New Jersey, consisting of 21 arches. PRR's renewed interest in masonry construction attracted attention in Britain, because for decades American railroads had relied largely upon iron and steel bridges, while masonry construction for main-line bridges in Britain and elsewhere had been common and widespread. To this effect, *The Railway Magazine* wrote:

> *The use of [stone] in place of steel marks a distinct change in the attitude of the road toward this kind of structure. American engineers and railroad men generally had come to believe that the days of expensive stone bridges for railroads had passed, but as an officer of the Pennsylvania said, recently this company has reached the conclusion that there is real economy in them, and hereafter it is like that most the bridges to be built by it will be of stone.*

The advantages of well-built masonry viaducts are straightforward. Because they are so solidly built, they eliminate need for speed restrictions and require minimal maintenance under normal conditions. Also, their ability to withstand great loads

At 2:45 p.m. on October 13, 1997—Columbus Day—Amtrak train No. 40, the *Three Rivers*, works eastward on the former Pennsylvania Railroad Middle Division between Granville and Lewistown. The train is crossing over the Juniata River on a stone arch bridge built by the PRR during its late-nineteenth-century upgrading and reinforced in modern times with vertical and lateral concrete supports. Older reinforcements consisting of steel diagonal supports can be seen on the arches. Also note the three masonry piers—remnants of the earlier bridge replaced by the stone arches. *Brian Solomon*

means they are unlikely to require replacement despite the continued increase of axle weights and train speeds. They are also less likely to suffer from washouts. At a time when there had been great concern about bridge failures, particularly from badly constructed iron trusses, investment in stone bridges undoubtedly sat well with the public and investors.

PRR's extensive application of masonry bridges and viaducts was one of the most prominent examples of such construction, but by no means the only one, as Baltimore & Ohio, Fitchburg Railroad (later part of the Boston & Maine system), the New Haven, New York Central, and Philadelphia & Reading also made considerable use of masonry arches for replacement spans in the later nineteenth century.

Among the most prominent bridges on the PRR was its crossing of the Susquehanna, a few miles north of Harrisburg, Pennsylvania, at Rockville. In the 1840s, this river was viewed as one of the most formidable obstacles to the new railroad being built to connect Philadelphia and Harrisburg. Rockville was selected because the Susquehanna was shallow there and a relatively low bridge could be built at this location without the need for tall supports or movable spans. Edwin P. Alexander, writing 100 years after the fact in *The Pennsylvania Railroad: A Pictorial History* (1947), explained that the pier work began in 1847 but was delayed when the initial construction contract was abandoned. Between 1848 and 1849, a 23-span Howe truss was erected by Daniel Stone.

The completed Rockville Bridge was a showcase structure that served as a significant endorsement for the Howe design. However, when the railroad doubled its Main Line, the single-track bridge proved a significant bottleneck and in 1876–1877, was replaced with a double-track iron-lattice deck truss that Alexander states was 3,680 feet long and consisted of 21 spans 156 feet long and two spans 150 feet long, with each span consisting of three deck trusses. Gradually, the railroad increased line capacity by installing four main tracks over much of its route. This project was largely completed by 1896. By 1900, the double-track Rockville Bridge was again deemed inadequate; the railroad replaced it with what is regarded as the world's largest stone arch bridge, consisting of 48 arch spans. Alexander's specifications for the third Rockville Bridge are 3,820 feet long and 52 feet wide, with each arch spanning 70 feet and rising 46 feet above normal river level. It carried four main tracks, reduced to three during the Conrail era. This impressive structure survives today, serving

The Bow Ridge Bridges at Tunnelton, Pennsylvania, were both built by the Pennsylvania Railroad. The stone arch bridge was built in 1909 as part of a rebuilding of the original West Penn Railroad. The heavily built, rivet-connected, Warren deck-truss bridge was built in 1947 as part of a U.S. Army Corps of Engineering flood control project on the Conemaugh River. The bridge shortened what is now the Norfolk Southern Conemaugh Line by 0.8 miles and raised the line 19 feet, 6 inches higher than the 1909 line. *Patrick Yough*

Norfolk Southern on its heavily traveled east-west main line.

During 1866–1867, the Connecting Railway was built in Philadelphia to give Pennsylvania Railroad a through rail route between that city and the New York area. (Until 1910, when Penn Station opened, PRR's tracks terminated on the New Jersey side of the Hudson, opposite Lower Manhattan.) This new double-track Connecting Railway bridged the Schuylkill just to the north of the present-day Girard Avenue Bridge. The December 19, 1913, *Railway Age Gazette* described PRR's bridge as a combination of seven 60-foot brick arches faced with sandstone ashlar (smooth square or rectangular stones laid with mortar in horizontal courses) and a 242–foot-long central iron-deck truss. Repairs were necessary in 1873 to correct for flaws in the masonry caused by excessive thrust from the truss and related settling. By 1901, the iron span was no longer capable of supporting the weight of trains, so a new 242-foot steel span was built to replace it.

Despite the improved bridge, traffic levels demanded further improvement within the decade. Initially, the railroad anticipated replacing the truss span with a concrete arch and enlarging the width of the bridge for another two tracks. However after 1913, PRR's engineering department decided to build a new structure with space for five tracks.

For almost 130 years, this stone arch at Athol, Massachusetts, has remained in daily service. Above the keystone are the initials for the Fitchburg Railroad, later a component of Boston & Maine's primary east-west main line, today operated by Pan Am Railways. *Brian Solomon*

The Fitchburg Railroad built a number of well-made stone arches along its lines in Massachusetts and New Hampshire in the late nineteenth century. This compact arch may be considered little more than a culvert by some standards, but has carried the railroad over a stream at Erving, Massachusetts, for at least 120 years. Since it was built, this arch has been reinforced by vertical steel beams screwed in place across the spandrels. In front of it is a disused industrial spur running over a small girder bridge supported by reinforced concrete abutments. *Brian Solomon*

A Chicago & North Western autorack train glides across the Shopiere Bridge, a masonry arch at Tiffany, Wisconsin, on June 24, 1995. This unusual five-span bridge was built in the 1860s and sometimes is cited as the oldest stone arch bridge in the Midwest. Several varieties of stone from different parts of Wisconsin were used in its construction, resulting in its distinctive stratified appearance. *Brian Solomon*

This later bridge is a multiple-span concrete arch faced with sandstone that resembles the earlier structure. It was built in 1914 and was part of a larger improvement program increasing capacity in the Philadelphia area.

PLATE-GIRDER BRIDGES

The plate-girder bridge, like a number of other important railway innovations, appears to have originated in Britain. The greater malleability and high tensile strength of wrought iron as compared with cast iron caught the attention of British bridge designers. The new material could be riveted together easily, making onsite construction easier. H. Shirley Smith, author of *The World's Great Bridges* (1953), indicates the first wrought-iron girder bridge was built in 1832 near Glasgow, Scotland. George Stephenson was among the first to adapt iron girder bridges for railroad application. His son, Robert, also built several famous and very distinctive iron girder bridges, among them the application of a bowstring girder type for the High Level Bridge over the River Tyne between Newcastle and Gateshead, built between 1846 and 1849. Built at the same time was Robert Stephenson's Britannia Tubular Bridge across the Menai Strait, along with a similar, but shorter bridge, at Conwy, both located in Wales. The Britannia and Conwy bridges were fully enclosed through-girder designs, with the tracks laid through the "tubes" that made up the structural girder. Two tubes were required, one to carry each track.

Completed in 1850, the Britannia spanned 460 feet. It was supported by abutments on opposite sides of the Menai Straits and in the middle by three masonry towers. At the time of its completion, it

The Rockville Bridge is considered the longest continuous stone arch railroad bridge in the world. Built in 1902 to accommodate four tracks, it is the third bridge at this location. Piers for one of Pennsylvania Railroad's earlier bridges can be seen in the river to the right of the current structure. On May 4, 2002, a little more than 100 years after the stone arch bridge entered traffic, a westward Norfolk Southern double-stack container train eases across the famous structure. *William D. Middleton*

Westbound Conrail Mail-11 crosses the famous Rockville Bridge on October 25, 1984. The bridge spans the Susquehanna River between Rockville and Marysville, Pennsylvania. *Doug Eisele*

In 1914, this massive arch over the Schuylkill River in Philadelphia replaced the original 1867 double-track bridge made of stone arches and a metal truss span. Today, it carries Amtrak's Northeast Corridor. Although it resembles the stone arches it replaced, it is actually a reinforced concrete arch faced with sandstone. *Brian Solomon*

On the afternoon of October 29, 2006, AEM-7 electric 952 leads an Amtrak train bound for 30th Street in Philadelphia and Washington, D.C., crossing the Schuylkill River on the massive sandstone-faced arch viaduct completed in 1914. Beyond is the modern Girard Avenue Bridge that carries SEPTA's No. 15 trolley. *Brian Solomon*

In August 1979, a Chicago & North Western Fairbanks-Morse H10-44 switcher hauls a few gondolas across an unusual type of plate-girder bridge. Located at Oconto Falls, Wisconsin, it consisted of both deck and through spans resting on iron piers filled with concrete. This train operated five days per week between Stiles Junction and Oconto Falls, Wisconsin, and was among the last to use C&NW's F-M diesels. *Terry Norton*

was deemed the longest railway bridge in the world. In 1859, Stephenson built the Victoria Bridge at Montreal that used tubular construction similar to the Britannia.

In America during the mid-nineteenth century, the relatively high cost of metal and low cost of labor made plate-girder construction, as employed by Stephenson and others, prohibitively expensive for most railroad bridges. Plate-girder construction requires a larger amount of metal than do truss spans of the same length. By the 1890s, the situation had changed. Development of the steel industry had greatly reduced the cost of rolled-plate steel. A further improvement was the development of cheap riveting processes that enabled plates to be riveted on-site. While structural principles were similar, American steel-plate-girder bridges were unlike the early British bridges. Instead of enormous custom-designed structures such as the Britannia Bridge, American plate-girder construction was to standard prefabricated plans and typically used steel rather than wrought iron. Two basic styles of plate-girder construction have been widely adapted for railroad use: through- and deck-bridge designs. As with truss spans, in a through bridge, tracks run between the girder supports, and on a deck bridge the tracks are on top.

American railroad steel-plate-girder bridges are fairly simple in comparison to the various types of metal trusses. In *Bridge Engineering* (1916), J. A. L Waddell offered this basic description: "A plate-girder is a beam composed of a wide, thin plate, called the web, along each edge of

Canadian National's *Southwestern Local*, named for Halifax & Southwestern Railway, works over the bridge at Martins River, Nova Scotia, May 22, 1991, on CN's line between Halifax and Yarmouth. The line closed about a month after this photo was taken. The unusual bridge was typical of those on the line, consisting of timber pile trestle approaches and lightweight plate-girder deck spans. The latter were carried by masonry piers that previously supported an earlier bridge on this site. Montreal Locomotive Works' road switchers were built as models RS-18, which used B-B trucks, but were rebuilt as CN MR14a/MR14b, de-rated from 1,800 to 1,400 horsepower and equipped with A1A trucks (six-wheel trucks with the central axle unpowered) from retired RSC-13s and RSC-24s to lower the axle load–a necessity for work on light branch lines with bridges such as this. *Thomas L. Carver*

which there is riveted a flange that may consist of a variety of structural shapes."

The flanges are necessary to distribute horizontal compression and tensile forces to the cross-sections of the girder. Since these forces, called "moments," are greatest at the center of the span, it is often necessary to rivet several layers of plate together to form the top flange. These additional plates attached to the top and bottom flanges are called cover plates. Moments are, in effect, "bending" forces that must be resisted by the girders. In the web, stresses are greatest at the ends of the span. These "shear forces" are resisted within the girders by vertical ribs called stiffeners.

One primary advantage of plate-girder bridges is that they can be assembled off-site, then brought in by rail and installed using a minimum of time and labor. As the cost of labor rose following World War I, the higher material costs of plate-girder construction were offset by lower overall labor cost, making plate-girder bridges not only more sturdy and reliable, but more cost-effective, too. Simplicity, ease of installation, and lower costs have led railroads to use plate-girder bridges to replace a variety of older bridges, including many of the quirky nineteenth-century trusses of nominal length. As a result, during the first years of the twentieth century, plate-girder bridges became a predominant replacement span.

Plate-girder spans have been used cost-effectively for lengths up to 80 to 120 feet. Longer structures require multiple consecutive spans sharing intermediate piers for support. Plate-girder construction requires considerably more metal than a truss span of comparable length. Greater amounts of metal are not only more costly, but substantially heavier, which leads to less overall capacity for the trains themselves. Beyond lengths of 120 feet, truss spans are usually more cost-effective replacement solutions.

Plate-girder spans often have been used in conjunction with other types of construction. In addition to application as short-span girder bridges over roads, streams, and ponds, multiple-span plate-girder bridges supported by steel frame towers or masonry piers have been used widely to build long and tall viaducts. Where there are no concerns for the minimum clearance below rail-level, deck-plate girder spans are suitable, whereas either through-plate or through-truss spans are better choices where minimum below-rail clearance presents restrictions. Often, through-girder spans have been used for road crossings, both to carry the railroad over a road, and the road over a railroad, where minimum vertical under-clearance is a concern.

Canadian Pacific's line across the top of Maine was among the last main-line steam operations in the United States, surviving until 1960. CPR's eastward mixed train, known as "the Scoot," makes its way across the Onawa Trestle, located west of Brownville Junction, on April 9, 1955. Although technically a viaduct consisting of plate-girder and truss-deck spans supported by reinforced concrete piers, this bridge retains its colloquial name despite strict formal definitions excluding it from "trestledom." *Jim Shaughnessy*

This view of Robert Stephenson's Conwy bridge shows the westward abutments. Each of the twin iron tubes serves as a girder and carries a main-line track within. Despite the relatively modern appearance of the bridge tubes, this design dates to 1849. *Brian Solomon*

Robert Stephenson's famous tubular bridge at Conwy, Wales, remains in service today. The abutments on either side of the iron tubes were designed to emulate the thirteenth-century castle located at the west side of the bridge. Each of the iron tubes is 400 feet long. The Britannia Bridge, which also used the tubular design, was built by Stephenson across the Menai Strait several miles to the west of this bridge, but was destroyed by fire. On March 31, 2007, a Virgin Voyager passenger train is seen at the west end of the Conwy Bridge next to the famous castle.
Brian Solomon

CHAPTER 6

Concrete Arches

Concrete is neither a recent discovery nor new to bridge builders. Its wide-scale application, however, for railroad spans is primarily a twentieth-century phenomenon. When finally adopted, concrete was used for some of the most ambitious American railroad projects ever undertaken. In the formative era of concrete railroad bridge construction, these structures were viewed with awe and wonder. Today, the largest of these monuments still capture the enthusiasm of anyone who appreciates superb engineering.

Roman engineers employed concrete made with natural hydraulic cement in the creation of their magnificent viaducts and aqueducts, many of which remain standing today. American bridge builders experimented with cement in the first half of the nineteenth century: In 1818 natural cement was applied in the construction of aqueducts for New York's Erie Canal. Thirty years later, cement was employed in the construction of Erie Railroad's splendid limestone Starrucca Viaduct. Yet, these early

Left: The ghostly form looming above the village of Nicholson, Pennsylvania, on the evening of October 10, 2003, is the famous Tunkhannock Viaduct, lit by the rising moon. The bridge is a reminder of the economic power once held by America's railroads. The anthracite mined from the mountains in eastern Pennsylvania was burned long ago, while profits from transporting it were invested in infrastructure such as this bridge. *Brian Solomon*

At 8:35 a.m. on May 18, 1990, a Union Pacific freight led by C30-7 No. 2474 rolls out of a tunnel and across the massive North Fork Bridge on the former Western Pacific, deep in the North Fork Feather River Canyon. This open spandrel concrete arch bridge was built in the early 1960s as part of a massive line relocation in conjunction with the construction of Oroville Dam (1961 to 1968), which flooded the lower canyon and with it WP's traditional right of way. *Brian Solomon*

Concrete, which was successfully employed in bridge construction by Roman engineers more than 2,000 years ago, enjoyed a renaissance in the twentieth century. However, where Roman concrete aqueducts and other structures have stood for centuries, many more modern structures have begun to crumble within just decades. On January 19, 1956, New York Central E8A No. 4082 leads an eastward passenger train across a concrete overpass at the railway station in Rome, New York. *Jim Shaughnessy*

applications are but footnotes in the bigger picture. Widespread commercial use of cement came about as a result of the development of artificially produced concrete and steel-reinforcing techniques.

The first recorded reinforced concrete bridge in America was a 20-foot arched span installed in San Francisco's Golden Gate Park in 1889 by Austrian engineer Joseph Melan, a prolific bridge builder who built concrete arches incorporating rolled-steel beams or lattice arch ribs for support, many of them in the United States.

In the first decade of the twentieth century, this aesthetically pleasing variety of concrete arch gained favor with American bridge builders, who used it for a number of significant railroad bridges. Although American engineers were more hesitant to embrace reinforced concrete than were their European counterparts, several very significant railroad projects made prominent use of the new material in the first decade after 1900.

LACKAWANNA RECONSTRUCTION

In 1899, William H. Truesdale assumed the presidency of the Delaware, Lackawanna & Western, a prosperous anthracite coal road that connected Hoboken, New Jersey (opposite Manhattan), with

Although it hasn't carried a train in almost 30 years, Paulins Kill Viaduct—located in rural northwestern New Jersey—survives as a monument to the Lackawanna and its superb engineering. *Brian Solomon*

The abandoned former Lackawanna bridge over Delaware River at Slateford Junction, Pennsylvania, as seen on March 11, 2007. It consists of seven open-spandrel arches of various lengths and two small closed-spandrel arches. As built, it measured 1,450 feet in length. This was the smallest of the four large open-spandrel concrete-arch viaducts designed by DL&W bridge engineer Abraham B. Cohen for the railroad's early-twentieth-century improvements. It may carry trains again someday. *Brian Solomon*

Buffalo, New York, by way of Scranton, Pennsylvania. Truesdale brought a visionary approach to railroading and valued high-quality engineering. Under Truesdale, Lackawanna engineers used the latest technology of the time to rebuild the railroad into one of the finest lines in America.

According to Thomas Townsend Taber in *The Delaware, Lackawanna & Western Railroad, Part Two*, the railroad's first concrete bridge was a 40-foot span at appropriately named, Bridgeville, New Jersey, where a new station, also of concrete, was built in 1903. This project set the tone for Lackawanna's next major endeavor: Slateford Cutoff (also known as the New Jersey Cutoff and the Hopatcong-Slateford Junction line). While other railroads had embarked upon ambitious projects to reduce curvature and to even grades, nothing else quite compared with Lackawanna's new cutoff. It began in 1908 as a showcase for reinforced concrete construction. Running from a junction near Port Morris, New Jersey, 28.5 miles to Slateford Junction, Pennsylvania, this route was completely grade-separated with a largely tangent profile, requiring two large bridges, as well as substantial cuts and fills, 73 concrete culverts, and other short concrete spans. It was 11.1 miles shorter than Lackawanna's existing main line. The two largest bridges were both of the open-spandrel arch design. Historian William S. Young, who has pioneered much of the research on DL&W's bridges, emphasized in an interview (and his own writings) that DL&W's bridge engineer, Abraham B. Cohen, was the man largely responsible for the railroad's concrete arch bridge design, although Cohen was rarely credited for his work at the time.

Although a massive concrete bridge in its own right, Lackawanna's Martin's Creek Viaduct has been overshadowed by the larger Tunkhannock Viaduct located only a few miles railroad-direction east on the same line. The Martin's Creek span is 775 feet shorter and 65 feet lower than Tunkhannock, yet it is the second-largest bridge on the old Lackawanna. On October 11, 2001, CP Rail SD40-2 No. 5866 leads train 164 across the elegant open-spandrel arch bridge. Today, the route is operated by Canadian Pacific. *Joe Geronimo*

The seven-span Paulins Kill Viaduct was 1,100 feet long and 117 feet tall at its highest point, and it required an estimated 43,212 cubic yards of concrete and 735 tons of steel. Slateford Viaduct across the Delaware River at the west end of the cutoff, consisting of seven open-spandrel arches of various lengths and two small closed-spandrel arches, measured 1,450 feet long and about 64 feet above high water.

The success of Lackawanna's Slateford Cutoff was a prelude to its boldest venture: the epic rebuilding of its main line over Clarks Summit via Nicholson, Pennsylvania, to reduce substantially the grade and curvature of its difficult route between Scranton and Binghamton, New York. The all-new line, known as the Summit Cutoff (later as the Nicholson Cutoff), was 39.6 miles long and cut the main line by 3.6 miles, reduced the maximum gradient from 1.23 percent (uncompensated for curvature) to just 0.68 percent (compensated), and eliminated an estimated 2,400 degrees of curvature. Low grades require less power to move trains, while reduced curvature allows trains to operate at higher speeds. Like the Slateford Cutoff, the Summit Cutoff displayed the use of concrete at every turn. Stations and interlocking towers, tunnel faces, culverts, and, of course, bridges were all made of reinforced concrete. The project was closely followed by the trade media. Detailed articles appeared in *Railway Age Gazette*. The estimated cost of Summit Cutoff is often cited at $12 million—an enormous sum in pre–World War I times.

The cutoff's star attraction is the truly massive 12-span Tunkhannock Creek Viaduct at Nicholson. The bridge required 167,000 cubic yards of concrete, almost four times the amount used in the Paulins Kill Viaduct. With a rise of 242 feet from the stream to the top of the bridge, the concrete arches soar 2,375 feet across the valley and tower above wood-frame houses below. The ten principal spans are open-spandrel arches, while smaller closed-spandrel arches (known as abutment spans) are at either end of the bridge. Based on drawings and specifications published in the January 5, 1915, *Railway Age Gazette*, the main arches use a semicircular design with a 90-foot radius, giving each a 180-foot span. The outside radius for the top of the arch is 111 feet. Atop each arch are 11 smaller spandrel arches, each spanning 13 feet, which support the ballasted deck. Upon completion, Tunkhannock was the world's largest concrete bridge.

Often overshadowed by the enormity of the Tunkhannock Viaduct is the similarly constructed Martin's Creek Viaduct, located just a few miles

Constructed between 1904 and 1913, Henry Flagler's railroad reached approximately 128 miles across shallow waters, hopping from island to island using long viaducts and bridges. Originally, construction called for 50-foot-span, closed-spandrel, reinforced concrete arches—at the time, the pinnacle of modern design. Only the Long Key Viaduct, pictured here, was built to the original specifications. *Author collection*

On September 25, 1988, an Illinois Central Gulf freight crosses the open-spandrel reinforced concrete arch bridge at Kankakee, Illinois. IC was among the pioneer American railways in reinforced concrete bridge construction. *Mike Abalos*

railroad-direction west of Nicholson. This bridge lacks the size and symmetry of Tunkhannock, but was the second-largest bridge on the railroad. It consists of seven large open-spandrel arches, each spanning 150 feet, along with several smaller arches for a total length of 1,600 feet and a maximum height of 150 feet. Where Tunkhannock was designed for double track, Martin's Creek was designed for three tracks.

Conrail, which inherited the old Lackawanna, abandoned the Slateford Cutoff; its magnificent bridges stand and can be viewed by passing traffic on Interstate 80. In 1980, Conrail sold the Binghamton–Scranton section of the former Lackawanna to the Delaware & Hudson. Today DL&W's Summit Cutoff and its magnificent Tunkhannock Viaduct remains an active main-line link in America's freight network, hosting trains of Canadian Pacific and Norfolk Southern. At present, there are discussions to reopen the Slateford Cutoff for use in a new passenger route from New York City to Scranton.

FLORIDA EAST COAST KEY WEST EXTENSION

Among the most remarkable railroad lines ever constructed was Henry Flagler's Florida East Coast extension, which ran 128 miles across the islands and waterways of the Florida Keys to Key West. Like Truesdale, FEC president Flagler was a visionary. His extraordinary railroad line was truly a product of the time in which it was built—only in the early twentieth century could politics, economics, and technology enable private funds to bring to fruition such a fantastic project. Unlike the Lackawanna, which was experiencing tremendous growth of heavy freight and passenger traffic, FEC's line was purely a speculative venture. Today, an expensive, privately built railroad line across the water to an isolated terminal with virtually no heavy traffic would seem a bizarre undertaking, but at the time it made sense—to Flagler anyway. The October

Norfolk Southern symbol freight 16T (Birmingham, Alabama, to Allentown, Pennsylvania) glides across the former Reading Company arch bridge over the Susquehanna at Harrisburg, Pennsylvania, on July 7, 2003. Each of the 46 semicircular reinforced concrete arches spans 66 feet. The arches encase the old masonry piers of the 1890–1891 deck-truss bridge, which remained in service during the building of the new structure between 1920 and 1922. *Joe Geronimo*

In October 1963, a Reading Company T-1 4-8-4 leads one of the railroad's famous *Rambles* across the Susquehanna River Bridge at Harrisburg, Pennsylvania. This 3,507-foot-9-inch-long, 46-arch concrete structure on the Reading's Philadelphia, Harrisburg & Pittsburgh branch (today operated as Norfolk Southern's Lurgan Branch) replaced a 23-span deck truss built by Coffrod & Saylor in 1890 and 1891. The arched bridge is on a 0.7 percent grade, with the eastern abutment 54.65 feet above river level, compared with 79.15 feet for the western abutment. *Richard Jay Solomon*

6, 1905, *Railroad Gazette* set the tone when construction began:

> *The present southern seaport terminus of the Florida East Coast is at Miami, from where the Peninsular & Occidental Steamship Co.'s boats run to Nassau, Key West and Havana. Since the war with Spain, Cuba has experienced a tremendous commercial development and as a winter resort it has also become very popular. For these two reasons the new line, which promises to afford much better facilities for communication between Cuba and the mainland, seems to justify the hopes of its promoter as to future traffic.*

In *Landmarks of the Iron Road*, Middleton notes that by the time of Flagler's project, the Panama Canal was underway. A new deepwater port at Key West would get FEC closer to the Canal than any other railroad in the United States.

While sometimes this near-mythic railway line has been portrayed as one very long bridge, in reality, it was a number of viaducts and causeways connecting islands in the Florida Keys. Reinforced concrete was initially considered for much of the project, in part, because it was thought to withstand the highly corrosive forces of saltwater better than plate steel. In 1905, a standard form for a semicircular closed-spandrel arch spanning 50 feet was printed in *Railway Gazette*. This was used in construction of the Long Key Viaduct, a nearly 2-mile-long structure consisting of 180 arches. Distance from the mean water level to the inside crown of the arch was 25 feet, and it was 30 feet from water to rail level.

Dan Gallagher explained in *Florida's Great Ocean Railway* (2003) that FEC planned to reach the Long Key Viaduct with an extended causeway. This proved untenable, and when the section opened in 1908 the approach to Long Key consisted of a 1,512-foot-long wooden pile trestle. In 1913, the trestle was replaced by 35 concrete arches. As it turned out, Long Key was the only major portion of the Key West extension to use the 50-foot concrete arch plan. Subsequent spans used truss bridges, long plate-girder viaducts on concrete piers, and concrete arches of a more modest design. Gallagher wrote that the Pacet Channel Viaduct, the western-most portion of the Knight's Key Bridge, consisted of 210 closed-spandrel concrete arches, each spanning 35 feet. Rail level was an estimated 17 feet above the mean water line. Another of the long concrete arch viaducts was a portion of the Spanish Harbor Bridge, using 35-foot span arches for a bridge 3,312 feet long. Gallagher explains that concrete arches were used in areas of shallow water, while deeper locations were better

On April 14, 2002, Union Pacific DASH8-40CW 9556 leads a freight out of Eugene, Oregon, and across a modern bridge constructed from pre-stressed concrete girders and reinforced concrete piers. Modern railroad bridges often use the same materials and techniques used in highway bridge construction. *Brian Jennison*

suited to concrete-pier-supported plate-girder and through-truss spans.

Different types of bridge construction were determined by various economic and engineering parameters. The engineers in charge were Joseph Meredith and his assistant, William J. Krome; the latter assumed the role of chief engineer upon Meredith's untimely death in 1909. The project was nearly completed in January 1912, and on January 22, Henry Flagler rode the first passenger train across his railroad to Key West, although work on the bridges was still ongoing and continued until 1913, the same year that Flagler died, having enjoyed the physical manifestation of his vision, though his dreams for traffic on the Key extension never panned out. During its 22 years of operation the line was a sparsely traveled curiosity, and in September 1935 a disastrous hurricane badly damaged it. Afterward, Florida East Coast, already in bankruptcy, petitioned the Interstate Commerce Commission for abandonment of the extension. Much of the route was rebuilt into the Overseas Highway, opened to the public in 1938. Many of the concrete arch bridges survive in highway service today.

CONCRETE SPANS

Although the early concrete bridges of Lackawanna and Florida East Coast have long captured the imaginations of writers and historians, many

This monumental crossing of the James River at Richmond, Virginia, was built with 18 reinforced concrete arches. The bridge is 2,278 feet long. Its main spans are 116 feet across, while shore-based spans are 122 feet. Unlike the Lackawanna viaducts of similar reinforced concrete open-spandrel arch construction, the James River Bridge doesn't feature high-parapets on the sides. Originally, it served Atlantic Coast Line trains, later those of Seaboard Coast Line, and today is used by CSX and Amtrak. A pair of General Electric–built AC6000CWs leads a southward CSX freight across the bridge on February 16, 2004. *Brian Solomon*

other American railroads have also made good use of concrete. Kansas City Southern, Illinois Central, Pennsylvania Railroad, and Richmond, Fredericksburg & Potomac were among the lines that built impressive open-spandrel arched bridges.

Between 1920 and 1922, Philadelphia & Reading (Reading Company) rebuilt its Susquehanna River crossing at Harrisburg, Pennsylvania, with a 46-span closed-spandrel arch bridge. This project was described in a series of contemporary articles by Philadelphia & Reading's engineers in *Railway Age*. Officially designated by the railroad as Bridge No. 8, the new bridge was built by encasing the masonry piers used to support the old single-track, 23-span, wrought-iron through-truss bridge. The old bridge remained in service as the new one was built alongside it. The new bridge reached 3,507 feet across the Susquehanna and was built as a double-track structure using reinforced concrete arches, each spanning 66 feet.

CHAPTER 7

Movable Spans

Where railroad lines intersect navigable waterways, U.S. government regulations (traditionally under jurisdiction of the former War Department, but today under the Coast Guard) mandate various minimum clearances for channel width and overhead obstruction. Channel width dictates minimum span requirements, while overhead clearances mandate other considerations that influence the type of bridge to be built. In some instances, fixed railroad bridges can be built tall enough to provide adequate vertical clearance, however, sometimes the construction of long approach viaducts to bring tracks to the height necessary to clear the minimum vertical requirement is either cost-prohibitive or impractical.

The difficulties and high cost of operating railroads on grades, not to mention the great cost of building long and tall bridges, have discouraged railroads from fixed-span solutions; instead, railroads have more often addressed navigable waterway crossings by installing movable spans. In the

Right; The former Pennsylvania Railroad swing bridge on the High Line from New York Penn Station to Newark crosses the Hackensack River near Kearny, New Jersey. This heavily traveled double-track line serves hundreds of Amtrak and New Jersey Transit trains daily, keeping this bridge at the center of attention. Known as Portal Bridge, it is one of seven movable railroad bridges over the Hackensack. At 7:37 p.m. on August 6, 1997, NJ Transit train No. 3645 glides over Portal on its way to Matawan. *Patrick Yough*

4423

This former Illinois Central bridge across the Mississippi River at Dubuque, Iowa, uses a swing span to allow passage of river traffic. The bridge tender works from the shack on the swing span–seen here in the open position. At the time of this August 12, 1994, photograph, the line was operated by Chicago Central & Pacific, an Illinois Central Gulf spin-off that was reacquired by Illinois Central in 1996. Today, Canadian National owns the IC. John Stover, in his *History of the Illinois Central Railroad*, indicates that the original structure was a seven-span bridge that opened in 1869. *Brian Solomon*

United States there are an unusually large number of places where railroad lines cross navigable waterways. This has resulted in a large variety of movable-span bridges, more so than in most other nations. The greatest concentrations of movable spans are in the Northeast—particularly along the northern New Jersey coast and coastal New England—along with the Great Lakes region, the Mississippi River, the Gulf Coast, the San Francisco Bay Area, and the Pacific Northwest. Commonly known as drawbridges—a term often used interchangeably with movable span—the three most common types used by North American railroads are swing, bascule-lift, and vertical-lift bridges. Each has been built with a number of variations.

SWING BRIDGES

The swing bridge is the simplest and traditionally the most common type of movable span in railroad use. The type is often used for moderate-length spans and built using both truss and plate-girder construction. A swing bridge rotates laterally around a vertical pivot point. In its most common application, the swing bridge is designed with equal arms on either side of the pivot point, which eliminates the need for counterweights and complex engineering. In rare instances, swing bridges are constructed with arms of unequal lengths. These cases require careful counterbalancing.

There are three basic types of swing spans as defined by their method of bearing: center-bearing, rim-bearing, and combination, an amalgam of the first two types. All were described in detail in 1910, in the *Minutes of Proceedings of the Institution of Civil Engineers*, by James Athersmith Orrell: The center-bearing swing bridge essentially consists of primary longitudinal support girders supported by a central lateral girder that pivots around a central transverse girder, which transmits the entire dead load of the bridge when it is rotating. The bridge is cantilevered outward from the pivot girder.

In the rim-bearing design, primary support girders are balanced across a centrally placed circular drum, typically the same width as the span. Weight transfer to the bridge pier is accomplished using a ring of conical rollers on which the span rotates.

In the combination system, both center-bearing and rim-bearing equipment is used, each carrying a portion of the load.

Orrell noted that center-bearing swing bridge designs were most suitable for single-track spans of 200 feet or less, while larger spans were better suited to rim-bearing or combination designs. John Alexander Low Waddell, professional bridge designer, self-promotion artist, and prolific writer, commented in his 1916 book, *Bridge Engineering*,

New Jersey Transit train No. 1032, running as a push-pull from Montclair Heights to Hoboken, crosses Bridge No. 7.57 over the Passaic River at West Arlington, New Jersey, on September 17, 2002. This is a classic example of a plate-girder, rim-bearing swing bridge. Notice the ring of wheels atop the circular pier on which the swing span rests. Constructed in 1897, major improvements were implemented to this structure in 1911, with the last major modifications made in the 1950s, when New Jersey Route 21 was built under the west end. *Patrick Yough*

regarding the types of swing span draw bridges, that "In general it may be stated that while the rim-bearing draws are often more rigid than stable, the center bearing draws move with less friction."

The pivot point must be capable of supporting the dead load of the bridge during rotation, but it is not necessarily used to support the live load of a train crossing the bridge. In some swing bridge designs, once the bridge has rotated into the closed position and its ends are secured, the full weight of the span rests on end piers through a complicated system of moveable steel wedges driven through a transmission system. These wedges become the bearings for such bridges while closed. This is the dominant system for active swing bridges in America today. Typically, when the bridge is closed, the ends are locked to the abutments or end piers, which is necessary both to support the span and to keep the bridge from being damaged by a crossing train. The locks are integrated into the interlocking system at each bridge. If the bridge is not secured, the repeated pounding action caused by a live load at the loose ends will damage the span. Securing the ends is also important to minimize gaps between rail heads, since even very small gaps caused by a misaligned swing span may cause an accident.

A swing bridge offers several advantages over the other common types of draw bridges. In general, it requires less complicated engineering. It provides two openings that allow greater capacity for ships passing in both directions. Its opening has unlimited overhead clearance, which is an advantage where ships with tall masts require access. A swing can be easily situated on a skewed crossing of a river, estuary, or canal. And small swing spans can be operated manually.

While the swing bridge has remained the most popular type of movable span, it suffers from a

Milwaukee Road's Pontoon Bridges

By John Gruber

Floating, or pontoon, bridges, a uniquely Milwaukee Road solution to the high costs of iron and steel spans, served the railroad well on secondary lines for more than 80 years. The "pontoons" were actually wooden boats, with structures on them supporting the rails.

The railroad operated three bridges on the Mississippi River–at Wabasha, Minnesota, and two at Marquette, Iowa–and one on the Missouri River–at Chamberlain, South Dakota. These were not the only pontoon bridges, but they became well known, long-established institutions–symbols of the Midwestern line's independent spirit.

The Vermont & Canada (leased to the Central Vermont in 1884) and the Northern of New York built what apparently was the first pontoon railway bridge in 1848 to 1850. It connected the two lines at the outlet of Lake Champlain at Rouses Point, New York, and served until April 1, 1868. But its design problems were not as difficult as those faced on the

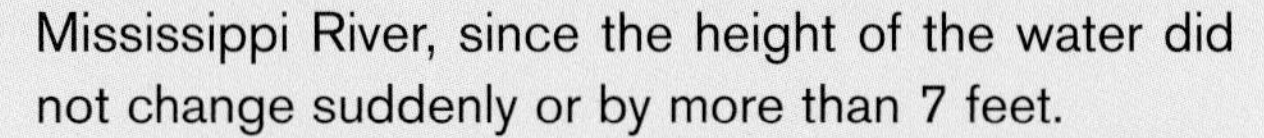

Mississippi River, since the height of the water did not change suddenly or by more than 7 feet.

John Lawler's 1874 patent for "floating draw bridges" set the pattern for the three Milwaukee Road crossings and for a bridge on the Panama Railroad. In 1867, Lawler, Milwaukee Road station agent at Prairie du Chien, Wisconsin, negotiated a contract (and later a franchise) to transfer freight across the Mississippi between Prairie du Chien and Marquette. After trying out steamboats, transfer barges, and car ferries, he built a pile bridge and placed barges in the openings for steamboats, but had problems getting trains on and off the barges.

In 1873, he hired Michael Spettel, a German shipbuilder who had moved to Prairie du Chien, at $60 a month to solve the problem, which included a 22-foot variation between high and low water levels. Spettel worked for months in a shop in Prairie du Chien, preparing a model with penknives. Lawler tried to change Spettel's plan, but was unsuccessful and only delayed construction. Spettel's plan was followed, and on April 15, 1874, the first train successfully crossed the river on the new pontoons (two were required because there were channels on the Wisconsin and Iowa sides of the river). The patent, filed by Lawler in his own name, was granted August 11, 1874. An article about it appeared in the *Transactions of the American Society of Civil Engineers* in 1884. Although the railroad had taken over operation of the bridge, Lawler retained the franchise and earned $1 a car. He died in 1891, and his family sold the franchise to the Milwaukee in 1894.

From the beginning, operation required a complex structure for stability. Since the water level on the river changed and the pontoon rides lower in the water when under load, the structure also had to overcome the

A Milwaukee Road SD9 leads a short freight over the railroad's last active pontoon bridge at Prairie du Chien, Wisconsin, across the Mississippi River, shortly before the bridge was abandoned. *John Gruber*

A period hand-tinted color postcard portrays Milwaukee's pontoon bridge over the Mississippi in the open position to allow passage of a riverboat and raft of logs. *Author collection*

arrangement approved by the ICC in 1952.

In "Floating Swing Spans for Railroad Bridges" in 1932, *Railroad Engineering and Maintenance* wrote about the limitations of the pontoon: "It imposes restrictions on the speed of trains and its movement in opening and closing the waterway is somewhat slower that that of movable bridge spans but where these disadvantages impose no serious restriction on rail and water traffic their lower cost has warranted their continued use on the two branch lines of the Milwaukee."

variations between track levels on fixed bridge trestles and the pontoons.

To open for boats, the pontoons had a pivot at one end and swung open and closed, like a door on a hinge, by winches that pulled a chain. The railroad trade press editorialized about the advantages and disadvantages of the pontoon. "The original cost, as well as the repairs, is a small fraction of what would be required for an iron structure," *Railway Review* said in 1892.

When the Milwaukee Road built its Black Hills Division west to Rapid City, South Dakota, it crossed the Missouri River at Chamberlain on a 366-foot pontoon. After 13 years of severe service, it developed leaks in spring 1918, and had to be replaced by second-hand girders to form a temporary 50-foot channel opening. A year later, two 300-foot through truss spans purchased from the Kansas City Terminal provided a permanent bridge.

The Milwaukee's longest pontoon span, 396 feet, was built in 1882 at the Mississippi River crossing between Wabasha and Trevino, Wisconsin, on a line to Chippewa Falls, Wisconsin. Replacement spans were built in 1890 and 1932. It remained in service until approach spans washed out in high water in 1951. Rather than repair the spans, Milwaukee negotiated trackage rights through Winona to Trevino, an

At Prairie du Chien, new pontoons were launched for the east channel (209 feet long) in 1914 and for the west channel in 1916 (276 feet long). In 1927, the railroad took the east channel span out of service for repairs, replacing it with a temporary pile trestle during the winter. Similar repairs were made to the west channel pontoon when the navigation season closed in 1932–1933.

Until 1960, the Milwaukee operated daily passenger service using the Prairie du Chien–Marquette pontoons. Soon after the train was taken off, the railroad requested approval from the Interstate Commerce Commission to abandon service across the river. Seventeen employees, working in three shifts, were needed to operate the pontoons. Electricity had been used for the last six years to open and close the bridge, but steam power still raised and lowered the tracks. At a hearing at Prairie du Chien, the railroad said costs for the crossing were $96,000 a year and estimated that $1,054,785 would be needed in the next five to six years to repair the two pontoons and key portions of the trestles.

The inevitable closing came in 1961 as the ICC granted permission to abandon the spans and the 1.5 miles of track included in the river crossing, bringing an end to the colorful and labor-intensive pontoon era.

The former Davenport, Rock Island & North Western Railway's Mississippi River Bridge uses a central swing span, which, like many swing bridges, is operated by a bridge tender stationed on the bridge. Opening and closing a swing bridge is generally more time consuming than with a vertical lift, in part because of greater difficulty in realigning the tracks. In the foreground is the machinery used to separate the tracks prior to opening the span and to rejoin them again when it has been closed. Today, this bridge is used several times a day by BNSF Railway trains and those from the Iowa, Chicago & Eastern. *Brian Solomon*

This low plate-girder deck on CSX's River Subdivision in New Jersey crosses Bellman Creek, a tributary of the Hackensack River. It features one short span that previously served as a small bascule lift, a remnant of the once heavily industrialized waterways that make up the New Jersey Meadows. In March 2000, northward CSX autorack train Q271 crosses it just south of "CP5" in North Bergen, New Jersey. *Tim Doherty*

variety of disadvantages that sometimes result in a preference for the bascule or lift bridge. Waddell, a proponent of bascule and lift bridge designs, was quick to highlight the many failings of swing spans. By virtue of its design, a swing bridge must occupy the middle of the channel, which makes navigating the channel more hazardous, especially in narrow channels. The bridge also requires considerable lateral space to enable it to turn, and a swing span generally requires more time to open and up to three times as much time as other movable spans to become fully closed, properly aligned, and capable of handling trains. If a railroad wishes to increase line capacity, a swing bridge must be entirely replaced because it is not practical to place swing spans parallel to one another. Finally, because of its bearing surface, a swing span is expensive to maintain.

BASCULE BRIDGES

Waddell traced the history of the bascule movable span, writing that the name derives from French for "balance," which describes the essence of its design. In its basic form, the bascule draw is a platform rotating on a lateral hinge. The French were the first to adapt the type in the nineteenth century. The first large-scale railroad application of the bascule in America is believed to be in 1894, when the Metropolitan Elevated Railroad bridged the Chicago Canal using a Scherzer rolling lift bridge (see below).

There are two basic bascule arrangements, two common bridge types, and several different designs based on each type. All require heavy counterweights balanced against the span. Sometimes known as leaf bridges, bascules can be arranged as single-leaf or facing double-leafs,

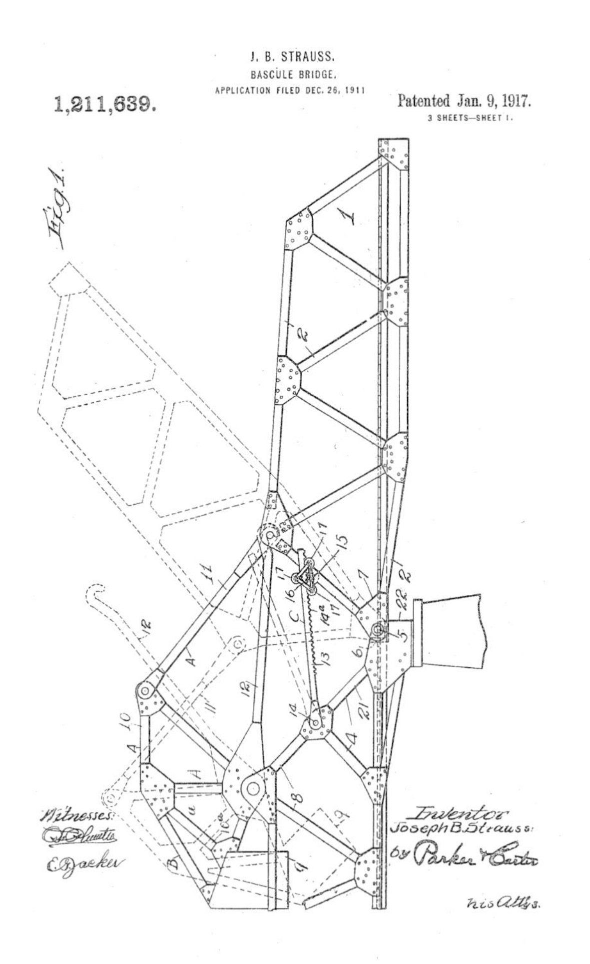

Joseph B. Strauss designed some of the most popular types of bascule lift bridges for railroads. This drawing is from his 1917 patent. Strauss is most famous for the Golden Gate Bridge. *U.S. Patent No 1,211,639, January 9, 1917*

An Erie Railroad Alco RS-3 leads a commuter train across the Strauss heel trunnion-type bascule bridge near Newark, New Jersey. This line has since been closed to traffic. *Richard Jay Solomon*

wherein the opening ends of the spans face one another. Since aligning the tracks on opposing spans proves problematic, the majority of railroad bascules are of the single-leaf arrangement.

The most common bascules used in railroad applications are the trunnion type and the rolling lift type. The basic trunnion is the easiest to understand. The bridge and counterweight are balanced on a fixed lateral axis located at or very near the center of gravity. The counterweight allows the bridge to be opened with a minimum of energy. As the counterweight is lowered the span is raised. However, basic trunnion bridges are fairly rare for railroad applications. Among the most common type of trunnion bascules used by railroads is the Strauss Heel trunnion, designed by Joseph B. Strauss in 1908 and patented in 1910. (Strauss was granted a second bascule patent in 1917.) This is more complex than the basic type described above. Here, the counterweight and span are hinged on separate but parallel axes and linked using a moveable steel framework in the shape of a parallelogram. The bridge is opened using a fixed pinion gear that engages a rack attached to the steel frame. As the pinion rotates, it adjusts the framework, compressing the form of the parallelogram and in so doing lowering the counterweight and lifting the span. Among its advantages, it minimizes the space required for the counterweight and lifting mechanism, and it allows a larger maximum angle when the bridge is fully open, which in turn allows for greater waterway clearance.

The rolling lift bascule is aptly described by its name. The bridge opens by rolling back on fixed tracks attached to its piers. Where the trunnion bridge opens with a fixed center of gravity, on the rolling bascule the center of gravity moves as the span is opened. The Scherzer and Rall are the most common varieties. With the Scherzer,

Continued on page 150

One of several heavy rolling bascule lift bridges at South Station, Boston, Massachusetts, as photographed on November 23, 1988. This angle clearly shows the semicircular rear quadrants and track that support the span when lifting. Above the quadrants are the counterweights *Brian Solomon*

16
17
18
19

Above: The former New Haven railroad bridge at Westport is a complex structure. In addition to accommodating the four main-line tracks and dual lifts, its towers support the 11,000-volt alternating current catenary and high-voltage feeders used to power the trains. Birds rest on the catenary supports and related steelwork in between passing trains and bridge lifts on a brisk November 2002 afternoon. *Brian Solomon*

Opposite: During 1904–1905, New Haven Railroad rebuilt many of the bridges along its main line between New York and New Haven as part of a capacity improvement that increased the line from two to four main tracks. At the Mianus River crossing in Cos Cob, Connecticut, New Haven's earlier structure consisted of seven fixed Whipple deck trusses and a center-swing span 235 feet long. Portions of the old masonry were retained during the upgrading; the original swing span was replaced by this pair of Scherzer rolling bascule lift bridges. Today, Metro-North, which operates and maintains the line owned by the State of Connecticut, designates the spans at Cos Cob as Bridge 29.90. On July 16, 2006, the dual draws were open to allow passage of a yacht. *Patrick Yough*

Following: Metro-North's Bridge 44.32 over the Saugatuck River at Westport, Connecticut, uses another pair of parallel Scherzer rolling bascule lifts that date from the 1904–1905 line expansion project. The plate-girder deck lift spans are each 98 feet long; the total length of bridge 44.32 is 458 feet. Hundreds of daily Metro-North commuter trains destined to and from Grand Central Terminal in New York use the four-track line, as do dozens of Amtrak long-distance passenger trains and the infrequent Providence & Worcester freight. *Patrick Yough*

BNSF's M-PASBAR (Pasco, Washington, to Barstow, California) symbol freight starts across the former Oregon Trunk Bridge over the Columbia River between Wishram, Washington, and Celilo, Oregon, on September 24, 2004. The lead locomotives are framed in the massive vertical lift that is in the up position to clear river traffic. Since the Burlington Northern–Santa Fe merger of 1995 and the Union Pacific–Southern Pacific merger of 1996, traffic on the old Oregon Trunk Route has soared and the line has become much busier. *Tom Kline*

Below: Sunset over the Columbia River on August 9, 2003, finds BNSF's northward M-BARPAS (Barstow, California, to Pasco, Washington) rolling across the former Oregon Trunk Bridge between Celilo, Oregon, and Wishram, Washington. *Brian Jennison*

On January 13, 1990, an oil tanker is moored in the Natches River Bay at Beaumont, Texas, near the vertical lift bridge used at that time by Southern Pacific and Kansas City Southern. This bridge was a replacement span installed in place of a gigantic rolling bascule. *Tom Kline*

Continued from page 144
the bridge rolls back on its axis, supported by a pair of large, fixed curved surfaces called quadrants, located at the rear of the structure. As with the Strauss, a fixed pinion gear rotates the structure back by engaging a lateral rack. With the Rall system, as the span opens, a roller supports the bridge on a specially built girder track. A rotating pinion gear separate from the main structure pulls the span open by engaging a pivoting movable rack attached to the bridge.

Because they are balanced, bascule bridges are very efficient to operate and open more quickly than swing bridges. While they use less lateral space, they require more vertical clearance and, depending on the variation, may need space for the counterweight. Among the disadvantages of bascules, as highlighted by Waddell, is that they are difficult to design for skewed crossings and can be difficult to operate in very windy conditions.

VERTICAL LIFT

Vertical lift bridges are typically straightforward designs. Towers on either side of the channel provide both a lifting mechanism, using counterweights, and support for the bridge as it is lifted. A notable feature of lift bridges, by nature of their design, is that the counterweights are equal to the weight of the span, in contrast to bascule designs, which require heavier counterweights. The vertical lift is the "newest" type of movable span and peculiar to American operations. Waddell, a pioneer of the modern vertical lift, having designed the South Halsted Street span in Chicago in 1894, credits truss designer Squire Whipple, who built short lift spans across the Erie Canal as early as 1872, as among the first to apply lift bridges to movable spans. Having emerged in the late nineteenth century, the vertical lift became a dominant form of the large movable span in the twentieth. The Pennsylvania Railroad built several lift bridges in the Northeast in the

The Sarah Mildred Long Bridge spans the Piscataqua River between Kittery, Maine, and Portsmouth, New Hampshire. The top deck carries U.S. Highway 1a; the bottom deck carries the old Eastern Route of the Boston & Maine, now operated by Pan Am Railways. While the highway is heavily traveled, the railroad sees infrequent trains to the Portsmouth Naval Base. In this January 26, 2007, photo, the vertical lift span is up to allow passage of a ship. *Candace Pitarys*

Louisiana & Delta is a short-line spin-off that operates several former Southern Pacific branch lines in Louisiana. On January 22, 1996, L&D CF7 1504 leads a short freight across this simple short-span lift bridge at Delcambre, Louisiana. Notice the counterweights atop of the structure. *Brian Solomon*

At 4:21 p.m. on August 19, 2003, CSX symbol freight Q-433 (Selkirk Yard, New York, to Oak Island Yard, New Jersey) works westward across Bridge 3.15 over Newark Bay on the former Pennsylvania Railroad Greenville Branch. Commonly known as the Upper Bay Bridge, this vertical lift span forms a vital link in the freight network that serves the New York metro area, but in a capacity quite different than originally planned. PRR built this route across Newark Bay between 1901 and 1904, shortly after it acquired control of the Long Island Rail Road. At that time, Greenville Yard was envisioned as a gateway to Long Island freight traffic via a tunnel that was never built, but from time to time is still proposed. *Patrick Yough*

Opposite: Among the advantages of lift bridges are their ability to span large channels and to be built side by side, such as these drawbridges over the Hackensack River between Jersey City and Kearny, New Jersey. On February 18, 2001, CSX Q-409 (Selkirk, New York, to Waycross, Georgia) eases over the former Pennsylvania Railroad P&H Branch lift bridge at "HACK." P&H stands for "Passaic & Harsimus." *Tim Doherty*

early part of the twentieth century, and in 1935 New Haven broke all previous records, building its 544-foot vertical lift bridge across the Cape Cod Canal at Buzzards Bay, Massachusetts.

Vertical lift bridges are well suited for long span crossings, multiple-track decks, skewed arrangements, and double-deck bridges, where a road is used on one level and the railroad on the other. Among the other advantages of the lift type is that it can be built with any variety of roadbed or deck, including heavy ballasted decks. Lift bridges can be opened more quickly than swing bridges and because of their design, align tracks automatically when closing. Among their disadvantages are limited and defined vertical clearances and the necessity for tall towers. They are also very expensive to build, even when compared to other moveable spans. Where vertical clearances may be limiting, neither bascule nor vertical lift bridges are appropriate.

CSX

Above: New Haven's vertical lift bridge over the Cape Cod Canal at Buzzards Bay, Massachusetts, was the longest of its kind when completed in 1935. It spans 544 feet and can be lifted to 130 feet above the water. New Haven cabooses mark the ends of their respective freights that have just met at the passing sidings railroad-west of the bridge. *Jim Shaughnessy*

Left: In April 1980, three Burlington Northern GP9s and a single F7A lead a freight on the former Great Northern line along the shores of Puget Sound at Steilacoom, Washington. The train is crossing an uncommon type of lift bridge patented by Joseph B. Strauss in 1911. Instead of the more common cable lift arrangement, this design uses dual pairs of pivoted counterweight arms on fixed supports at each end of the bridge to raise and lower the lift span. Better known for his heel trunnion-style bascule lift bridge–also patented in 1911–Strauss' masterpiece was San Francisco's Golden Gate Bridge. *Thomas L. Carver*

Sources

BOOKS & MONOGRAPHS

Alexander, Edwin P. *The Pennsylvania Railroad: A Pictorial History.* New York: W. W. Norton, 1947.

Ashman, Robert. *Central New England Railroad.* Salisbury, Conn.: Salisbury Association, 1972.

Baker, B. *Long Span Railroad Bridges.* Philadelphia: Baird, 1870.

Bell, J. Snowdon. *The Early Motive Power of the Baltimore and Ohio Railroad.* New York: Angus Sinclair Company, 1912.

Binney, Marcus, and David Pearce, eds. *Railway Architecture.* London: Bloomsbury Books, 1979.

Boller, Alfred P. *Practical Treatise on the Construction of Iron Highway Bridges.* New York: John Wiley & Sons, 1876.

Brightmore, A. W. *Structural Engineering.* London: Cassell, 1908.

Bryant, Keith L. *History of the Atchison, Topeka and Santa Fe Railway.* New York: Macmillan, 1974.

Burgess, George H., and Miles C. Kennedy. *Centennial History of the Pennsylvania Railroad.* Philadelphia: Pennsylvania Railroad Company, 1949.

Campbell, Marius R., et al. *Guidebook of the Western United States*, 6 Vols.: *Northern Pacific*; *Overland*; *The Santa Fe*; *Shasta*; *Denver & Rio Grande*; *Southern Pacific.* Washington, D.C.: U. S. Geological Survey, 1915–1933.

Casey, Robert J., and W. A. S. Douglas. *The Lackawanna Story.* New York: McGraw-Hill, 1951.

Clarke, Thomas Curtis, et al. *The American Railway: Its Construction, Development, Management, and Appliances.* New York: Scribner's, 1889.

Condit, Carl. *Port of New York, Vols. 1 & 2.* Chicago: University of Chicago Press, 1980, 1981.

Corliss, Carlton J. *Main Line of Mid-America: The Story of the Illinois Central.* New York: Creative Age Press, 1950.

Cupper, Dan. *Horseshoe Heritage: The Story of a Great Railroad Landmark.* Halifax, Pa.: Withers Publishing, 1996.

Derrick, Peter. *Tunneling to the Future: The Story of the Great Subway Expansion That Saved New York.* New York: New York University Press, 2001.

Droege, John A. *Passenger Terminals and Trains.* New York: McGraw-Hill, 1916.

Encyclopedia Britannica, 1911 Ed. Vol. 4, pp. 534*ff.*

Gallagher, Dan. *Florida's Great Ocean Railway.* Sarasota, Fla.: Pineapple Press, 2003.

Giedion, Sigfried. *Space, Time and Architecture, 4th Ed.* Cambridge, Mass.: Harvard University Press, 1963.

Gillespie, W. M. *A Manual of the Principles and Practice of Road-Making, 4th Ed.* New York: A. S. Barnes, 1851.

Gordon, J. E. *The Science of Structures and Materials.* New York: Scientific American Library, 1988.

Graf, Bernhard. *Bridges That Changed the World.* Munich: Prestel, 2002.

Griffin, William E. *One Hundred Fifty Years of History along the Richmond, Fredericksburg, and Potomac Railroad.* Richmond, Va.: Richmond, Fredericksburg, & Potomac, 1984.

Harlow, Alvin F. *The Road of the Century.* New York: Creative Age Press, 1947.

———. *Steelways of New England.* New York: Creative Age Press, 1946.

Harris, W. H. *Central Electric Light and Power Stations and Street and Electric Railways.* Washington, D.C.: U. S. Department of Commerce, Bureau of the Census, 1912.

Haupt, Herman. *General Theory of Bridge Construction, 1st Ed.* New York: Appleton, 1855.

Hofsommer, Don L. *Southern Pacific, 1901–1985.* College Station, Texas: Texas A&M, 1986.

Hosking, William. *Essay and Practical Treatises on the Theory and Architecture of Bridges.* London, 1843.

Hungerford, Edward. *Men of Erie.* New York: Random House, 1946.

———. *Story of the Baltimore & Ohio.* New York: Putnam, 1928.

Karr, Ronald Dale. *The Rail Lines of Southern New England.* Pepperell, Mass.: Branch Line Press, 1995.

Kirby, Richard Shelton, and Philip Gustave Laurson. *The Early Years of Modern Civil Engineering.* London: Oxford University Press, 1932.

Kirkman, Marshall M. *Building and Repairing Railways.* New York: World Railway Publishing Company, 1903.

Klein, Maury. *Union Pacific, Vols. I & II.* New York: Doubleday, 1989.

Lamb, W. Kaye. *History of the Canadian Pacific Railway.* New York: Macmillan, 1977.

Lavis, Fred, and Maurice E. Griest. *Building the New Rapid Transit System of New York City.* New York: Engineering News, 1915.

Law, Henry. *The Rudiments of Civil Engineering.* London: Crosby Lockwood and Company, 1881.

Macdougall, Walter M. *The Old Somerset Railroad.* Camden, Maine: Down East Books, 2000.

Marshall, John. *The Guinness Book of Rail Facts and Feats.* Enfield, U.K.: Guinness Superlatives, 1975.

McCullough, David. *The Great Bridge.* New York: Simon & Schuster, 1972.

McGannon, Harold E. *The Making, Shaping and Treating of Steel, 8th Ed.* Pittsburgh, Pa.: United States Steel Corporation, 1964.

Merritt, Frederick S., et al. *Standard Handbook for Civil Engineers, 4th Ed.* New York: McGraw-Hill, 1995.

Middleton, William D. *Grand Central, the World's Greatest Railway Terminal.* San Marino, Calif.: Golden West Books, 1977.

———. *Landmarks on the Iron Road.* Bloomington: Indiana University Press, 1999.

———. *Manhattan Gateway: New York's Pennsylvania Station.* Waukesha, Wis.: Kalmbach Publications, 1996.

———. *Metropolitan Railways: Rapid Transit in America.* Bloomington, Ind.: Indiana University Press, 2003.

———. *When the Steam Railroads Electrified, 1st Ed.* Milwaukee, Wis.: Kalmbach Publications, 1974.

Moffat, Bruce G. *The "L": The Development of Chicago's Rapid Transit System, 1888–1932* (Bulletin 131). Chicago: Central Electric Railfan's Association, 1995.

Nock, O. S. *The Railway Heritage of Britain.* London: Joseph, 1983.

Petroski, Henry. *Engineer of Dreams.* New York: Vintage Books, 1995.

———. *To Engineer Is Human.* New York: Vintage Books, 1985.

Plowden, David. *Bridges: The Spans of North America.* New York: Norton, 1974, 2002.

Protheroe, Ernest. *The Railways of the World.* London: Routledge, 1914.

Raymond, William G. *Elements of Railroad Engineering, 1st Ed.* New York: John Wiley & Sons, 1908.

Reeves, William F. *The First Elevated Railroads in Manhattan and The Bronx of the City of New York.* New York: New York Historical Society, 1935.

Reier, Sharon. *The Bridges of New York, 1st Ed.* New York: Quadrant Press, 1977.

Repairing and Strengthening of Old Steel Truss Bridges. New York: American Society of Civil Engineers, 1979.

Roberts, Charles S. *West End: B&O Cumberland to Grafton 1848–1991.*

Baltimore: Barnard, Roberts & Co., 1991.
Roth, Leland M. *Understanding Architecture: Its Elements, History and Meaning*. Boulder, Colo.: Westview Press, 1993.
Schleife, Hans-Werner, et al. *Metros der Welt*. Berlin: Trans-Press, 1992.
Seyfried, Vincent F. *The Long Island Rail Road: A Comprehensive History, Part 7: The Age of Electrification, 1901–1916*. Garden City, N.Y.: Seyfried, 1984.
Shank, William H. *Historic Bridges of Pennsylvania*. Manassas, Va.: Buchart-Horn, 1966.
Signor, John R. *Southern Pacific–Santa Fe–Tehachapi*. San Marino, Calif.: Golden West Books, 1983.
Sinclair, Angus. *Development of the Locomotive Engine, 1st Ed.* Cambridge, Mass.: MIT Press, 1907.
Smith, H. Shirley. *The World's Great Bridges*. London: Phoenix House, 1953.
Snell, J. B. *Early Railways*. London: Octopus Books, 1972.
Solomon, Brian. *The American Steam Locomotive*. Osceola, Wis.: MBI Publishing Company, 1998.
———. *Railroad Stations*. New York: Metro Books, 1998.
———. *Railway Masterpieces: Celebrating the World's Greatest Trains, Stations and Feats of Engineering*. Iola, Wis.: Krause, 2002.
Solomon, Brian, and Mike Schafer. *New York Central Railroad*. Osceola, Wis.: MBI Publishing Company, 1999.
Solomon, Richard J., and Kathleen M. Solomon. *Passenger-Psychological Dynamics*. New York: American Society of Civil Engineers, 1968.
Starr, John W., Jr. *One Hundred Years of American Railroading*. New York: Dodd, Mead, 1928.
Steinman, David B. *A Practical Treatise on Suspension Bridges: Their Design, Construction and Erection*. London: Chapman & Hall, 1929.
Steinman, David B., and Sara Ruth Watson. *Bridges and Their Builders*. New York: Dover, 1957.
Stilgoe, John R. *Metropolitan Corridor*. New Haven, Conn.: Yale University Press, 1983.
Stover, John F. *History of the Baltimore & Ohio Railroad*. West Lafayette, Ind.: Purdue University Press, 1987.
———. *History of the Illinois Central Railroad*. New York: Macmillan, 1975.
Straub, Hans. *A History of Civil Engineering: An Outline from Ancient to Modern Times*. Cambridge, Mass.: MIT Press, 1964.
Taber, Thomas Townsend, and Thomas Townsend Taber III. *The Delaware, Lackawanna & Western Railroad, Part One*. Muncy, Pa.: Thomas T. Taber III, 1980.
———. *The Delaware, Lackawanna & Western Railroad, Part Two, 1899–1960*. Muncy, Pa.: Thomas T. Taber, 1981.
Talbot, F. A. *Railway Wonders of the World, Vols. 1 & 2*. London: Cassell & Co., 1914.
Turneaure, Frederick, ed. *Cyclopedia of Civil Engineering*. Chicago: American Technical Society, 1912.
Turner, Gregg M., and Melancthon W. Jacobus. *Connecticut Railroads*. Hartford, Conn.: The Connecticut Historical Society, 1989.
Waddell, J.A.L. *Bridge Engineering*. New York: John Wiley & Sons, 1916.
Walker, J. B. *Fifty Years of Rapid Transit 1864–1917*. New York: Law Printing, 1918.
Waters, L. L. *Steel Trails to Santa Fe*. Lawrence, Kan.: University of Kansas Press, 1950.
Webb, Walter Loring. *The Economics of Railroad Construction*. New York: John Wiley & Sons, 1912.
———. *Railroad Engineering*. Chicago: American School of Correspondence, 1912.
Wellington, Arthur M. *The Economic Theory of the Location of Railways, 5th Ed.* New York: John Wiley & Sons, 1897.
Whipple, Squire. *A Work on Bridge Building: Consisting of Two Essays*. Utica, N.Y.: H. H. Curtiss, 1847.
———. *An Elementary and Practical Treatise on Bridge Building*. New York: D. Van Nostrand, 1873.
White, John H., Jr. *Early American Locomotives*. New York: Dover, 1979.
Wilson, Neill C., and Frank J. Taylor. *Southern Pacific: The Roaring Story of a Fighting Railroad*. New York: McGraw-Hill, 1952.
Winchester, Clarence. *Railway Wonders of the World, Vols. 1 & 2*. London: Amalgamated Press, 1935.
Yates, John A. *Standard Specifications for Railroad & Canal Construction*. Chicago: The Railway Age Publishing Company, 1886.
Young, William S. *Starrucca: The Bridge of Stone*. Published privately, 2000.

PERIODICALS, REPORTS, PAPERS, MINUTES, ET CETERA

1912 Street and Electric Railway Census.
American Railroad Journal and Mechanics' Magazine. Published in the 1830s and 1840s.
American Railway Engineering Association Proceedings & Manuals. Chicago.
Baldwin Locomotives. Philadelphia, Pa.: no longer published.
Buck, Sylvia G. "Some Bridge Builders in Warren [Mass.]," notes on unpublished research, January 28, 2000.
———. "William Howe, 1803–1852," notes on unpublished research.
Buckley, Tom. "Reporter at Large: The Eighth Bridge," *New Yorker*, Vol. 88, No. 48 (January 11, 1991), p. 56.
Classic Trains. Waukesha, Wis.
"Completing the New York Rapid Transit Subway," *Scientific American*, June 11, 1904, pp. 461*ff*.
"Floating Swing Spans for Railroad Bridges," *Railroad Engineering and Maintenance*, 1932.
Griggs, Francis E. "Squire Whipple—Father of Iron Bridges," *Journal of Bridge Engineering*, May/June 2002, pp. 146–156.
Huntington, Ellsworth. "The Water Barriers of New York City," *Geographical Review*, Vol. 2, No. 3. (September, 1916), pp. 169–183.
Institution of Civil Engineers. *Minutes of the Proceedings*. London.
Jane's World Railways. London.
Johnston, R. E. "New York Elevated Railroads," Paper No. 1704, Minutes of the Proceedings of the Institution of Civil Engineers (London), Vol. LXVII, Session 1881–2, Part II.
Journal of Bridge Engineering. New York: American Society of Civil Engineers.
Journal of the Irish Railway Record Society. Dublin.
Lindenthal, Gustav. *N. Y. Connecting R.R., Hell Gate Arch Bridge, General Plan of Structure as Built*. Blueprint No. 846, March 28, 1917.
Locomotive & Railway Preservation. Waukesha, Wis.: no longer published.
Modern Railways. Surrey, U.K.
Official Guide to the Railways. New York.
Parsons Brinckerhoff Hall & MacDonald. *Regional Rapid Transit: A Report to the San Francisco Bay Area Rapid Transit Commission*. New York, 1955.
Rail. Peterborough, U.K.
Railfan & Railroad Magazine. Newton, N.J.
RailNews. Waukesha, Wis.: no longer published.
Railroad History (formerly *Railway and Locomotive Historical Society Bulletin*). Boston, Mass.
Railway Age (variously *Railway Age Gazette*). Chicago and New York.

Railway Gazette. New York: 1870–1908.
Railway Magazine, The. London.
Railway Review (also *Railway and Engineering Review*). Chicago: 1876–1926.
Thomas Cook European Timetable. Peterborough, U.K.
Today's Railways. Sheffield, U.K.
Trains. Waukesha, Wis.
Transactions of the American Society of Civil Engineers. New York.
"Truss Bridges," *Quaboag Plantation*, Spring 1988.
Vintage Rails. Waukesha, Wis.: no longer published.

UNITED STATES PATENTS

5862X. Mar. 6, 1830. S. H. Long. Truss Bridge.
8743X. Apr. 3, 1836. Ithiel Town. Truss Bridge.
9340X. Jan. 23, 1836. S. H. Long. Truss Bridge.
1,192. June 24, 1839. Henry Wilton, Wrightsville, Pa. Construction of Bridges: Truss Bridge.
1,397. Nov. 7, 1839. Stephen H. Long, U. S. Army. Wooden-Framed Suspension Bridge: Truss Bridge.
1,398. Nov. 7, 1839. Stephen H. Long, U. S. Army and Marietta, Ga. Wooden-Framed Brace Bridge.
1,445. Dec. 27, 1839. Herman Haupt, York, Pa. Lattice (Truss) Bridge.
1,711. Aug. 3, 1840. W. Howe. Manner of Constructing the Truss-Frames of Bridges and Structures.
2,064. Apr. 24, 1841. Squire Whipple, Utica, N.Y. Construction of Iron Truss Bridges.
3,523. Apr. 4, 1844. Thos. W. Pratt, Norwich, Conn., and Caleb Pratt, Boston, Mass. Truss-Frame of Bridges.
4,004. Apr. 22, 1845. Geo. W. Thayer, Springfield, Mass. Wooden Bridge: Truss.
4,359. Jan. 15, 1846. Thomas Hassard, New York, N.Y. Truss Bridge.
4,945. Jan. 26, 1847. John A. Roebling, Pittsburgh, Pa. Suspension Bridge.
8,624. Jan. 6, 1852. Wendel Bollman, Baltimore, Md. Construction of Bridges: Suspension Bridge.
9,090. July 6, 1852. A. Bradway and E. Valentine, Monson, Mass. Construction of Bridges: Truss Bridge.
10,887. May 9, 1854. Albert Fink, Baltimore, Md. Truss Bridge.
16,446. Jan. 20, 1857. D. C. McCallum, Owego, N.Y. Truss Bridge.
16,728. Mar. 3, 1857. Albert Fink, Parkersburg, Va. Bridge-Truss.
154,055. Aug. 11, 1874. John Lawler, Prairie du Chien, Wis. Improvement in Floating Draw-Bridges.
277,039. May 8, 1883. Gustav Lindenthal, Pittsburg, Pa. Suspension Bridge.
311,338. Jan. 27, 1885. Gustav Lindenthal, Pittsburg, Pa. Arch-Bridge.
430,428. June 17, 1890. Gustav Lindenthal, Pittsburg, Pa. Suspension-Bridge.
506,571. Oct. 10, 1893. John A. L. Waddell, Kansas City, Mo. Lift-Bridge.
952,485. Mar. 22, 1910. John A. L. Waddell, Kansas City, Mo. Bascule Bridge.
974,538. Nov. 10, 1910. Joseph B. Strauss, Chicago, Ill. Bascule Bridge.
1,038,226. Sept. 10, 1912. Joseph B. Strauss, Chicago, Ill. (Lift) Bridge.
1,211,639. Jan. 9, 1917. Joseph B. Strauss, Chicago, Ill. Bascule Bridge.

INDEX